Anthropocene Reading

## AnthropoScene
### THE SLSA BOOK SERIES

**Lucinda Cole and Robert Markley, General Editors**

**Advisory Board:**
Stacy Alaimo (University of Texas at Arlington)
Ron Broglio (Arizona State University)
Carol Colatrella (Georgia Institute of Technology)
Heidi Hutner (Stony Brook University)
Stephanie LeMenager (University of Oregon)
Christopher Morris (University of Texas at Arlington)
Laura Otis (Emory University)
Will Potter (Washington, D.C.)
Ronald Schleifer (University of Oklahoma)
Susan Squier (Pennsylvania State University)
Rajani Sudan (Southern Methodist University)
Kari Weil (Wesleyan University)

Published in collaboration with the Society for Literature, Science, and the Arts, AnthropoScene presents books that examine relationships and points of intersection among the natural, biological, and applied sciences and the literary, visual, and performing arts. Books in the series promote new kinds of cross-disciplinary thinking arising from the idea that humans are changing the planet and its environments in radical and irreversible ways.

# Anthropocene Reading

Literary History in Geologic Times

Edited by
Tobias Menely and
Jesse Oak Taylor

The Pennsylvania
State University Press
University Park,
Pennsylvania

Library of Congress Cataloging-in-Publication Data

Names: Menely, Tobias, editor. | Taylor, Jesse O., editor.
Title: Anthropocene reading : literary history in geologic times / edited by Tobias Menely and Jesse Oak Taylor.
Description: University Park, Pennsylvania : The Pennsylvania State University Press, [2017] | Series: AnthropoScene: the SLSA book series | Includes bibliographical references and index.
Summary: "Considers the implications of the Anthropocene, the proposed geological epoch in which a human 'signature' appears in the lithostratigraphic record, for literary history and critical method. Explores the status of reading in the history of geology, and of geohistory in literature"—Provided by publisher.
Identifiers: LCCN 2017025913 | ISBN 9780271078724 (cloth : alk. paper) | ISBN 9780271078731 (pbk. : alk. paper)
Subjects: LCSH: Literature—History and criticism—Theory, etc. | Ecocriticism. | Geology in literature.
Classification: LCC PN441 .A73 2017 | DDC 809—dc23
LC record available at https://lccn.loc.gov/2017025913

Copyright © 2017
The Pennsylvania State University
All rights reserved
Printed in the United States of America
Published by The Pennsylvania State University Press,
University Park, PA 16802–1003

The Pennsylvania State University Press is a member of the Association of American University Presses.

It is the policy of The Pennsylvania State University Press to use acid-free paper. Publications on uncoated stock satisfy the minimum requirements of American National Standard for Information Sciences—Permanence of Paper for Printed Library Material, ANSI Z39.48–1992.

# Contents

Introduction | 1
Tobias Menely and Jesse Oak Taylor

1. Anarky | 25
Jeffrey Jerome Cohen

2. Enter Anthropocene, Circa 1610 | 43
Steve Mentz

3. The Anthropocene Reads Buffon;
or, Reading Like Geology | 59
Noah Heringman

4. Punctuating History Circa 1800:
The Air of *Jane Eyre* | 78
Thomas H. Ford

5. Romancing the Trace: Edward Hitchcock's
Speculative Ichnology | 96
Dana Luciano

6. Partial Readings: Thoreau's Studies as Natural
History's Casualties | 117
Juliana Chow

7. Scale as Form: Thomas Hardy's
Rocks and Stars | 132
Benjamin Morgan

8. Anthropocene Interruptions: Energy Recognition Scenes and the Global Cooling Myth | 150
Justin Neuman

9. Stratigraphy and Empire: *Waiting for the Barbarians*, Reading Under Duress | 167
Jennifer Wenzel

10. Reading Vulnerably: Indigeneity and the Scale of Harm | 184
Matt Hooley

11. Accelerated Reading: Fossil Fuels, Infowhelm, and Archival Life | 202
Derek Woods

12. Climate Change and the Struggle for Genre | 220
Stephanie LeMenager

13. Ungiving Time: Reading Lyric by the Light of the Anthropocene | 239
Anne-Lise François

List of Contributors | 259

Index | 261

# Introduction

*Tobias Menely*

*Jesse Oak Taylor*

In "The Storyteller," Walter Benjamin proposes that we "imagine the transformation of epic forms"—the heroic epic, the fairy tale, the proverb, the legend—as "occurring in rhythms comparable to those of the change that has come over the earth's surface in the course of thousands of centuries."[1] Benjamin compares the long durations of geomorphological alteration, the gradual movement of uplift and sedimentation, to the drift of literary genres across history: "There is hardly any other form of human communication that has taken shape more slowly, been lost more slowly" (147). This incremental shape-shifting suggests a further implication of Benjamin's analogy, related not to the scale of time but to its formal ordering, a likeness between lithic strata and the shaping power of narrative to organize time. Perhaps it is this formal intimacy with stone, this sense of the Earth as a primordial medium, that explains why, in Benjamin's view, stories so often extend "a ladder . . . downward to the interior of the earth" (157), attending even to the "mineral . . . the lowest stratum of created things" (161). An alertness to the lithic as a narrative medium reflects back on human self-conception. The storyteller sees human eventuality in the mineral world, "a natural prophecy of petrified, lifeless nature—a prophecy that applies to the historical world in which he himself lives" (161). In "The Storyteller," Benjamin contrasts the epic forms, which convey "distance"—the vastness of the Earth, the amplitude of time—with the modern novel, which addresses readers who expect immediacy, "information" up to the moment and close at hand. Literary form thus has internalized modernity's accelerated tempo and so no longer echoes the dilatory rhythms and extended durations that Benjamin associates with geological time.

Though the dialogue between literary criticism and the Earth sciences rarely achieves the esoteric grandeur of "The Storyteller," it is a persistent feature of both modes of inquiry. This is more than a matter of metaphorical traffic, as in the motif of the "stone book" or the critical propensity for describing texts as having topographical depths to be mined or rifts to be exposed. Stones, as Jeffrey Jerome Cohen has written, are a "spur to ceaseless story," "ancient allies in knowledge making," "material metaphor[s]."[2] Narrative expresses a basic human imperative to understand our place in a dynamic world of water, weather, and rock. "Storytelling," Bruno Latour observes, "is not a property of human language, but one of the many consequences of being thrown in a world that is, by itself, fully articulated and active."[3]

Geology has long defined itself as a scientific discipline through a simultaneous disavowal and incorporation of literary modes, especially romance, the narrative form concerned with marvelous phenomena: enigmatic ruins, vast time scales, absent causes.[4] In *The Sacred Theory of the Earth*, the Restoration physicotheologist Thomas Burnet—whom Stephen Jay Gould credits with conceptualizing Earth's past in terms of a "sequential narrative"[5]—claims that any reconstruction of planetary history will exceed empirical explanation and rely on imaginative ways of knowing. Some men, he writes, "distrust everything for a Fancy or Fiction that is not the dictate of Sense, or made out immediately to their Senses. Men of this Humour . . . call such Theories as these, Philosophick Romances." Yet, claims Burnet, "such Romances must all Theories of Nature . . . be."[6] Geology took shape as a modern discipline through its rejection of fancy and fiction, yet because it deals with scales of space and time unavailable to human experience, it has never altogether transcended its provenance in imaginative narrative forms.[7]

The literary dimensions of geology—a practice of *reading* stratigraphic inscriptions and *narrating* evocative, if improbable, stories—become even more pronounced in the Anthropocene, the proposed geological epoch in which humans, collectively, have come to rival "some of the great forces of Nature in [our] impact on the functioning of the Earth system."[8] In *The Earth After Us* (2008), Jan Zalasiewicz, the convener of the Anthropocene Working Group, dramatizes the challenges of interpreting the stratigraphic inscription left by humanity. In what amounts to a work of science fiction, he imagines a species of extraterrestrial stratigraphers arriving on Earth to reconstruct this epoch and understand its agent: "A storyteller arrives, one hundred million years from now, to tell the tale of the human species."[9] The alien scientists sift through layers of concrete and plastic, attempting to comprehend the species that left such traces behind. Their task will be "to find the message left by the human race"—an inscription "written in the strata"—and "then to decipher it" (118). Such reading and narrating, however, need not await geologists from another world. The Anthropocene is, after all, not only an epoch in Earth's geohistory defined by the shaping influence of human activity. It is also the epoch in which our singular species reads its transformative presence in the Earth's strata, reads *itself* in the rocks, and in doing so establishes new stories about its identity and this planet.

When Paul Crutzen and Eugene Stoermer proposed the term "Anthropocene" in 2000, they dated it to James Watt's 1784 patent on the double-acting steam engine.[10] This specificity coalesced a disparate set of causes and consequences into a widely recognizable act of technological innovation. In this version, the Anthropocene appears as a story of the unintended outcomes of human ingenuity, a Promethean tale. But there are other Anthropocenes as well. The Anthropocene Working Group has recommended that an Anthropocene associated with the postwar Great Acceleration be formalized as an epoch in the geological time scale (GTS), a designation that will ultimately require ratification by the International Union of Geological Sciences. The working group arrived at its recommendation after considering various "boundary events," including the Neolithic revolution, the Columbian exchange, fossil-fuel-powered industrialization, and nuclear weapons testing. Each start date redefines the narrative, its eponymous agent—the Anthropos as agriculturalist, conquistador, inventor, industrialist, capitalist, cyborg—and thus the shape and potential outcomes of the story. As the geographers Simon Lewis and Mark Maslin argue, "the event or date chosen as the inception of the Anthropocene will affect the stories people construct about the ongoing development of human societies."[11] In selecting a global boundary stratotype section and point (GSSP), or "golden spike"—a beginning, a spatiotemporal origin—geologists give narrative shape to history.

The Anthropocene, however, has never been simply a term of stratigraphic relevance. Indeed, we can think of no concept that has resonated so widely, so quickly, across the disciplines in academia and in the popular press. It has inspired interdisciplinary journals, numerous articles, symposia, monographs, and cover stories in the *Economist* and the *Guardian*. We believe that scholars are debating the Anthropocene not because it names a clear-cut epoch in which social and geological history come into alignment, but rather because its implications productively unsettle conventional disciplinary modes of inquiry. Critiques that the Anthropocene is merely a fashionable buzzword—or worse, a term that naturalizes capitalism, imperialism, or social inequality—may actually symptomatize the difficult intellectual burdens the Anthropocene imposes on us. In our view, the Anthropocene has inspired such intense debate, from the biophysical sciences to the humanities, because it identifies a problem, a problem of how emergent forms of causality, operating *across* sociohistorical and planetary systems, have come to be read in the Earth's strata and then conceptualized and communicated.

The Anthropocene is not an easy story to tell, especially for disciplines established within the "modern constitution" defined by the separation of social signs from natural facts.[12] As Dipesh Chakrabarty contends, anthropogenic climate change marks the point at which "the wall between human and natural history has been breached," demanding a wholesale reevaluation of the conceptual apparatus upon which the discipline of history is predicated.[13] This breach poses an equally profound challenge for the sciences, insofar as that epistemological "wall" preserved the divide between subject and object upon which objectivity, one's separation from what one studies, is based. As Latour observes, "the very notion of objectivity has been totally subverted by the presence of humans in the phenomena to be described" ("Agency," 2). Knowledge of nature comes to be inseparable from knowledge of social systems, and vice versa. Jason Moore calls this the problem of the "double internality": human social and economic forms at once *shape and are shaped by* "biological and geological conditions."[14] Lewis and Maslin, to offer one example, note that the 1610 Orbis spike, an atmospheric $CO_2$ dip precipitated by the depopulation of the Americas, converged with the emergence of Immanuel Wallerstein's capitalist "world system."[15]

Any definition of the Anthropocene identifies a point of entanglement between the Earth system and social systems, wherein varied forms of causality, from the imperatives of capital accumulation to the manner in which $CO_2$ absorbs infrared radiation, intersect. The Anthropocene Earth system, to put this another way, includes not just the hydrosphere, atmosphere, biosphere, and lithosphere, but also diverse economies and energy systems, societies and symbolic orders. In the Anthropocene, all scholars are called upon to become Earth system humanists, which involves thinking about how these systems interrelate with, internalize, and destabilize one another. Just as geologists are learning to account for sociohistorical causality and the rhetorical implications of stratigraphy, humanists are learning about the carbon cycle, ice-core sampling, and thermodynamics. Scholars across the disciplines are asking, in new ways, what it means to read history: to define an archive, to posit causality, to name a period or epoch, to narrate resonant stories about continuity and change.

*Anthropocene Reading: Literary History in Geologic Times* takes an avowedly disciplinary approach to this multidisciplinary problem, navigating two interconnected imperatives: to read the Anthropocene as a literary object and at the same time to recognize the Anthropocene as a geohistorical event that may unsettle our inherited practices of reading. The authors in this volume examine the Anthropocene as a narrative, investigating the rhetorical protocols informing lithostratigraphic practice and revealing the inherently fictional and yet epistemologically productive quality of any periodizing marker. Our aim, however, is not to deconstruct the Anthropocene, to unmask its inescapable rhetoricity, or to assert a disciplinary precedence vis-à-vis scientific truth claims. All of the contributors to this collection grapple with the Anthropocene as a historical *event*, a momentous phase transition in the Earth system that exceeds narrativization. The Anthropocene provides an opportunity for literary studies to test and transform its methods by examining how the symbolic domain might, or might not, index a historicity that exceeds the human social relation and encompasses planetary flows of energy and matter.

The "Anthropocene" is a newly resonant term for a long-standing problem in geology: the status of the current, and unfinished, epoch and of humankind's distinct place in it. This is no surprise, really, given that geology came of age during the Industrial Revolution, mapping strata in coal seams and railroad cuts. In *Epochs of Nature* (1778), Georges-Louis Leclerc, Comte de Buffon, identified the seventh planetary epoch, the current "time of man," in terms of the civilizational advancement promised by abundant fossil fuels. A few years later, James Hutton's *Theory of the Earth* presented the Earth system as a "machine" modeled, as Martin Rudwick has shown, on the coal-fueled Newcomen steam engine.[16] In 1854, the Welsh geologist Thomas Jenkyn termed the current epoch the "Anthropozoic," a designation adopted by Samuel Haughton in his *Manual of Geology* (1865). In the United States, George Perkins Marsh published *The Earth as Modified by Human Action* in 1874, revising his earlier *Man and Nature* (1864). Russian scientists used the term "Anthropocene" as early as 1922.[17] The proposal to formalize the Anthropocene as an official epoch in the GTS thus marks a moment of heightened self-reflexivity in the history of geology and Earth system science. Crutzen and Stoermer first proposed the new epoch in the year 2000. The Anthropocene is a millennial concept, a theory of (geo)historical crisis that has followed in the wake of the "end of history" that Francis Fukuyama proclaimed at the fall of the Berlin Wall in 1989, the same year that Bill McKibben described anthropogenic climate change as the "end of nature."[18] It turns out that the only thing that came to an end was the momentary illusion that "history" and "nature" could be conceptualized as separate.

In January 2016, members of the Anthropocene Working Group published an article in *Science* with the unambiguous title "The Anthropocene Is Functionally and Stratigraphically Distinct from the Holocene." The article lays out the conceptual criteria for the new designation: "Any formal recognition of an Anthropocene epoch in the geological time scale hinges on whether humans have changed the Earth System sufficiently to produce a stratigraphic signature in sediments and ice that is distinct from that of the Holocene epoch."[19] The authors posit a straightforward relation between a geophysical claim, that humans have altered the Earth system, and a stratigraphic claim, that such change leaves a "signature," a sign that enables a clear delineation between epochs by marking a scale shift in the geomorphic agency of a single species. And yet this changing and this inscribing are not the same. They act on different objects, the "Earth System" and "sediments and ice." The case for naming the Anthropocene is not presented as an analysis of the anthropogenic forcing of the Earth system, a potentially catastrophic crossing of "planetary boundaries," but is instead premised on a more narrowly semiotic claim about the clarity of a "signature" recorded in a lithostratigraphic archive.[20]

The authors of the *Science* article examine a number of candidates for a suitably clear and long-lasting signature, from "technofossils" and "geochemical" markers, such as pesticide residue, to concentrations of atmospheric carbon dioxide and the biostratigraphic signature left by increasing extinction rates. They also anticipated the recommendation made to the International Geological Congress in August 2016, which suggested that the Great Acceleration replace the Industrial Revolution as the most compelling lower boundary for the Anthropocene. "The most widespread and globally synchronous anthropogenic signal," they write, "is the fallout from nuclear weapons testing" (Waters et al., "The Anthropocene," aad26225). In a 2015 article, the working group had already suggested that a mid-twentieth-century lower boundary is "stratigraphically optimal" because the first nuclear bomb test in 1945, which left a clear layer of radiocarbon in the rock strata, is coincident with, if not causally related to, the more consequential, although less stratigraphically significant, Great Acceleration.[21] Hence, when contemplating the formalization of the epoch, the focus on anthropogenic intervention in the Earth system, which is to say the identification of a distinct mode of geohistorical causality, gives way to the question of identifying a synchronous, unambiguous, and long-lasting signature. Semiotic criteria take precedence over a geophysical account.

In an essay in the *Anthropocene Review*, Clive Hamilton polemically diagnoses this stratigraphic sleight of hand. Those who privilege the legibility of the signature are "fixated on the marker at the expense of what is marked."[22] He calls this fixation the "golden spike fetish": "an event in world history" is confused "with a historical marker for it." If, after all, the primary goal were to align sign and cause, the increased atmospheric concentration of $CO_2$ would be the obvious candidate, since it is the main driver of global climate change and the most significant manifestation of anthropogenic intervention in the Earth system. However, $CO_2$ emissions and atmospheric concentrations constitute an incremental, if accelerating, process, one without clearly demarcated boundary events but with complex social, economic, and technological causes. Moreover, the climate and sea level "signals" associated with increased greenhouse gases are "not yet . . . strongly expressed." Multiscalar, multicausal phenomena that cut across biogeochemical and sociohistorical domains do not necessarily leave clear-cut, localizable signatures.

The stratigraphic search for a "signature" that marks the emergence of the Anthropocene is a search for its definitive agent, the one who signs. The autograph of the "Anthropos" attests to its presence as a coherent entity, much the way a signature on a legal document attests to the identity of the person who affixes it. A signature, as Jacques Derrida explains, is a distinct form of inscription, one that serves to counteract the nonpresence of a speaker in written communication. Whereas in a spoken utterance the embodied presence of the speaker can be assumed, in writing the absence of a living person may be counteracted by the presence of a signature. The unique status of the signature derives from its clearly embodied origin. It attests to its author's "having been present," to an instance of "present punctuality," a specific person acting, and this action constitutes a singular spatiotemporal "event."[23] Yet to be meaningful, a signature must also endure; it must be "able to be detached from the present and singular intention of its production," legible even in the absence of its inscriber. A signature is a *supplement*, seeming to bear the "force" and "intention" of its inscriber in a form that survives the absence of its original source. Hence, the use of the signature as the trace or symptom of an elusive phenomenon, which has a long history in the sciences, takes on a newly appropriate resonance in the Anthropocene. The signature uniquely conveys the former presence of its author; it is a mark that remains, and remains meaningful, in the absence of the signatory. So, on one hand, the working group is attempting to establish, on the basis of a signature, the identity of an Anthropos as a single entity capable of acting as a planetary force. On the other hand, this signature must be commensurate with other stratigraphic markers: not only globally synchronous but also legible in the absence of other "historical" archives. This is why stratigraphers have objected to references to social history in identifying the epochal "boundary event," as in Lewis and Maslin's reference to the Nuclear Test Ban Treaty rather than the first appearance of nuclear residue.[24] For the stratigraphers, the signature must stand alone.

This tension between inscription and system is long-standing in geology, which depends on the constant negotiation between coconstitutive imperatives to delineate the Earth's strata—an enterprise often understood in semiotic terms, as an act of reading—and to account for the forces of planetary change: to *periodize* and to *historicize*.[25] The Anthropocene, however, introduces a new form of causality into the Earth system. Stratigraphers focus on a signature rather than an agent, in part because the actual status of the Anthropos poses problems they are not equipped to confront. The working group works backward, locating a legible signature and on that basis positing the existence of a species capable of altering the Earth system.

With this in mind, it should come as no surprise that the most resounding critique of the Anthropocene concept coming from the humanities focuses on the very question that, as we see it, the language of the "signature" attempts to forestall: the identity of the Anthropos. *Who*, precisely, leaves this signature, given that *Homo sapiens*, as a species, is defined by immense cultural and social variation? Responsibility and vulnerability are asymmetrically distributed in the changing Earth system. Given that the Anthropocene is decidedly not coextensive with the evolution of our species, but is rather an event that occurs in historical time, would it not be more precise to identify the Anthropocene with the distinct historical conditions in which human societies achieve geologic agency? Andreas Malm and Jason Moore have each suggested that "Capitalocene" better reflects the sociohistorical drivers of the new epoch. Malm argues that "this is the geology not of mankind, but of capital accumulation"; arising out of social conflict and exploitation, fossil capitalism is the "very negation of universal species-being."[26] Donna Haraway and Anna Tsing use "Plantationocene" to emphasize the epoch's inherently imperialist ecology and to make us "pay attention to the historical relocations of the substances of living and dying around the Earth as a necessary prerequisite to their extraction."[27]

These are immensely important critiques, both in foregrounding the ethicopolitical stakes of the Anthropocene and in focusing attention on the actual socioeconomic systems that constitute geologic agency. What such criticism of the Anthropos overlooks, however, is that for scientists the designation of a single species as an agent is a *specifying* move rather than a universalizing one. The point is not that *all* humans are transforming the Earth system but that a single species in the biosphere is transforming the planet, a significant event in geologic time. The working group attempts to elide the problem of the social through its invocation of a stratigraphic signatory, while a critical humanities perspective insists on social variation and relations of power but is often inattentive to the broader biogeophysical systems in which humans intervene as a distinct agent. The perceived incompatibility between these positions on the Anthropos attests finally to the deep epistemological challenge of conceptualizing the double internality.

The nomination of the Anthropocene is finally a stratigraphic prerogative, for it is stratigraphy that bequeaths us the geological time scale. The members of the working group are clear that, having accepted as axiomatic the significant human intervention in the Earth system, their job is to identify "a signature that is distinct from those of the Holocene and earlier epochs," to approach the designation of the Anthropocene in terms that "are consistent with those used to define other Quaternary stratigraphic units" (Waters et al., "The Anthropocene," aad26221). They also acknowledge that this is an impossible task. The question of whether to formalize the Anthropocene in the GTS, they write, is "a complex question, in part because, quite unlike other subdivisions of geological time, the implications of formalizing the Anthropocene reach well beyond the geological community. Not only would this represent the first instance of a new epoch having been witnessed firsthand by advanced human societies, it would be one stemming from the consequences of their own doing" (aad26228).

The Anthropocene is not only a break within the stratigraphic record but also an event that in effect breaks stratigraphic practice. As Bronislaw Szerszynski insightfully observes, such inscription works against our long-standing view of lithic impenetrability and in so doing disrupts the basis of the stratigraphic record itself: "The Anthropos will thus 'lie' in the strata in a different sense, in a different plane, not 'true'—as a perjurer, disrupting the semiotic logic of geology as much as its materiality."[28] In the context of the Anthropocene, stratigraphy's protocols of reading lead to questions about how human assemblages have come to constitute a planetary force of nature, questions that are only answerable outside of its disciplinary framework. Moreover, unlike other stratigraphic demarcations, which are ascribed retrospectively, the Anthropocene is unfinished, a tale without an ending. Indeed, the working group acknowledges as much at the end of an article in which they address the tension between Earth system science and stratigraphy directly, noting that geologic ages are ultimately determined not by the boundary events at their edges but by the stable climatic patterns that constitute the parameters of life within them. Citing the 2016 *Science* article, they write, "It is clear from both chronostratigraphic and Earth System perspectives that the Earth has entered the Anthropocene, and the mid-20th century is the most convincing start date."[29] However, in the very next sentence they go on to explain that "the Earth System is still in a phase of rapid change and the outcome is not yet clear." And they acknowledge, "The ultimate nature of the Anthropocene cannot yet be determined." In the midst of mounting their most forceful case to date for the formalization of the Anthropocene, these members of the Anthropocene Working Group allow that any characterization of the new epoch is provisional at best. The Anthropocene will ultimately be defined not by the point at which it began, but by the conditions of life within it. No matter where the GSSP is affixed, it will not determine the arc of the Anthropocene so much as the point at which the Holocene came to an end. Nonetheless, insofar as it will shape the stories we tell about human agency and human responsibility, the formalization of the Anthropocene will have material implications, potentially transforming the Anthropos itself.

The methodological predicaments we have been tracking in the stratigraphic discourse have been paralleled by a pronounced methodological disquiet in literary studies in the twenty-first century. Questions of method and rationale, of how and why we read literature, have always been a feature of literary studies, a broad-tent discipline that makes room for cultural critics and aesthetes, biographers and textual editors, empiricist historians and speculative theorists. The emergence of the Anthropocene as a multidisciplinary problem, however, has coincided with a malaise, and a new modesty, in literary studies. Literary scholars in the new millennium have been actively debating the legacy of theory, the future of method, and the coherence of literature as an object of study. We are asking how we justify the resources dedicated to our work—reading, teaching, and writing about literature—in an age of neoliberal austerity and STEM ascendance.

While no less rancorous, the theory decades (roughly 1970–2000) were defined by an unusual confidence in the purpose of our discipline. Knowledge, identity, and authority were understood to be constituted within a symbolic order that literary critics had powerful tools for unlocking. Groundbreaking works of literary history and criticism read works symptomatically, identifying the breaks in a text's legibility that express a broader psychic, linguistic, or social causality, whether the law of the father, the workings of *différance*, or the conflict over the means of production. In *The Political Unconscious* (1981), for example, Fredric Jameson trained readers to seek not a text's "unified meaning," as contained within its "organic form," but rather to pursue the "rifts and discontinuities within the work." These "clashing and contradictory elements" are "reunified" not in the text but in the critic's identification of a sociohistorical "process of production."[30]

In more recent years, a number of literary scholars have expressed a greater modesty, disavowing metalanguage and the ambitions of critical "suspicion," the impulse to demystify or destabilize in the act of reading.[31] In their introduction to a special issue of *Representations*, Sharon Marcus and Stephen Best advocate for a "surface reading" comparable to the "weakly" interpretive work of natural-historical classification, positioning literature scholars as more like stratigraphers than Earth system scientists.[32] Literary scholars working in the digital humanities have looked to quantitative methods to analyze "big data," establishing new archives and models, though it remains an open question whether such interpretive practices have produced compelling ways of rereading literary and cultural histories.[33] New formalists have turned to the organizing shapes and patterns that are shared by literary works and social systems, in what Caroline Levine dubs "strategic formalism."[34] New materialists and posthumanists have sought to establish methods of reading premised on a flat ontology, broadening our conception of signs, agents, and relations so as to resituate humans, and human meaning-making, in a broader constellation of beings.[35]

As an inherently global problem, the Anthropocene dovetails with the resurgence of interest in world literature and deep time in the work of scholars like David Damrosch, Wai Chee Dimock, and Franco Moretti.[36] Anthropocene reading shares this scholarship an attention to flows, trajectories, and systems that go beyond national borders and human time scales, while at the same time attending to the interplay of these systemic relations through fine-grained analysis. Like world literature, Anthropocene reading also depends on translation, not between languages but between disciplines. Ecocriticism, meanwhile, has long entered into dialogue with science. Initially characterized by a rejection of "theory" in favor of empirical realism drawn from biology, ecocriticism has shifted its focus to questions of social difference and environmental justice, exemplified in Rob Nixon's *Slow Violence and the Environmentalism of the Poor*, while other twenty-first-century work, such as Stephanie LeMenager's *Living Oil*, has been marked by an engagement with energy and matter.[37] Given ecocriticism's recourse to scientific principles, we might trace this arc in terms of the sciences it avows: first biology and ecology, then sociology and political science, and now stratigraphy and Earth system science. This process has been generally one of addition rather than substitution, as each wave intersects with and refracts those that preceded it.

In assembling this volume, we wondered whether the Anthropocene could clarify or complicate these debates about literary reading in the twenty-first century. The challenge of reading natural and social history in their double internality, which we identify in the stratigraphic discourse, takes inverse form in the humanities. Sociosymbolic phenomena have to be conceptualized in relation to the inhuman forms, forces, and scales of planetary systems and geologic time. We asked our contributors not only to read the Anthropocene but to consider how the Anthropocene might require us to read differently. What if the history implied by the dictum "always historicize" turns out to be not the internality of social relations but rather social relations as they shape and are shaped by thermodynamic, biospheric, atmospheric, and hydrological processes? Can we extend our own definitions of texts, signs, and traces? What can formalism do with nonhuman forms? Can literary reading provide insight into the Anthropocene's paradoxical alignment of precarity and agency, political urgency and deep time? Does literary history register modes of affect and experience related to thermodynamic, geological, and atmospheric processes? How can postcolonial and Indigenous studies be brought to bear on the question of species being? What might it mean to read geohistory symptomatically? Can the accelerated transformation of literary forms noted by Benjamin be understood to express broader patterns of change in energy production and the organization of biospheric systems? How might the Anthropocene inform current debates between historical materialists and new materialists, formalists and surface readers, historicists and post- or transhistoricists? How can we, as readers and critics, enter into dialogue with scientists without collapsing the differences between our disciplines: the differences between poiesis and physis, between poems and ice cores, textual and lithic archives, narrative and algorithmic ways of knowing?

We asked our contributors to articulate a method of Anthropocene reading and to show how it operated in practice. Something different, and deeply illuminating, happened instead. Rather than staging a consistent methodological practice, the contributors to this collection all read improvisationally, drawing on a range of conceptual tools, theories, and practices. It turns out that when your object of concern is something like the Anthropocene—multiform, multiscalar, multicausal, multitemporal—a commitment to methodological consistency may be exactly the wrong approach. In the pages of *Anthropocene Reading*, you will see psychoanalytic, philological, and deconstructive gestures. Our readers unpack metaphors and metonymies. They examine the affordances and limits of genre: allegory, romance, the medieval mystery play, the realist novel, experimental poetry. They stage experiential predicaments. They critique. They take up narratological problems: superpositioning, catastrophe, the vortex. They read forms, signs, fossils, structures, traces, symptoms. They tarry with the negative and hold out hope for messianic reversals. They get close to the text, down to the punctuation. This multiplicity of approaches leads us to the conclusion that the strength of our reading practices in literary studies may derive not from methodological rigor but from the acceptance of inconsistency, the belief in complexity, the attention to contradiction, and the labor of translation.

Reading in the Anthropocene is an invariably polyglot, salvage practice in which we employ all of our tools to discover meaning amid the ruins. Indeed, many of our readers emphasize the limits of knowledge and the inexpressible qualities of the Anthropocene: it becomes knowable only in incompletion or negation. There is a modesty in Anthropocene reading, but it is not the modesty of one who claims to merely describe. Our contributors universally accept that no single method can fully account for the various forms of Anthropocene causality and Anthropocene mediation. A number of them advocate methods that are defined by partiality and incompletion. They tend to practice forms of symptomatic reading, exploring textual depths and rifts, but without the invocation of a secure critical vantage point that would exempt the reading itself from the very condition it seeks to diagnose.

If it is not possible to enact a narrowly consistent Anthropocene reading practice, it is possible to perform individual readings that showcase the adaptability and innovative range of interpretive methods. Certain key problems in Anthropocene reading—the literary mediation of geohistory, the relation of literature to other (inhuman) media, narrative form and unconformity, the identity of the Anthropos, the formalization of scale variance and scale change—recur across the volume. Taken together, the chapters of this book evidence a practice of reading literary history in the context of geohistorical transition, a practice defined by a shared commitment to the interpretation of human and natural history in their double internality. No matter when (or if) the Anthropocene is formalized by the International Union of Geological Sciences, or what signature is ultimately vested with the authority to proclaim its coming of age, other dates and other markers will retain their significance because of the stories they tell. The instrumentalization of fire, the birth of agriculture, the first written word, the conquest of the Americas, the Industrial Revolution, the Great Acceleration: each offers a signal moment in the entanglement between human societies and the Earth system, an entanglement that is constitutive of human history.[38] The Anthropocene debate provides an occasion for, once again, recognizing this connection, for reckoning with the biogeochemical and thermodynamic contexts in which transpire all history and all symbolic activity. In this sense, the "Anthropocene" in our title does not promise a single practice or method, but rather establishes the conditions under which *all* reading must henceforth proceed.

Among the central preoccupations of this collection is the problem of periodization, dating. Stratigraphers invoke not a day or a decade as the time unit in which human history comes to suddenly intersect with geologic time, but a year: 1610, 1784, 1945.[39] As Steve Mentz observes in his contribution here, such dating "concentrates [the] mind": the "provisional" closure of the single year is a way of salvaging "form from inside disorder." A date imposes a division, establishing an end and a beginning. One order of things gives way; another takes hold. As inflection or flash point, the single date invokes the catastrophism of Cuvierian geology rather than the incremental shifts of Lyellian uniformitarianism.

Strikingly, the two chapters in this collection that focus on specific years—Mentz on 1610 and Tom Ford on 1800—also explore punctuation, the marks of syntactic closure and transition. This seemingly incongruous pairing, crystallized in the two meanings of "period," (a double-meaning shared by the term "epoch") underlines the scalar shifts that Anthropocene reading demands. Mentz observes, "[M]arks of punctuation that separate epochs and narratives from each other" can also be the basis for producing "genre hybridities" helping us to "create new things with old tools." Ford identifies an eccentric punctuation pattern in Charlotte Brontë's *Jane Eyre* (1847), the colon-dash (:—), which paradoxically conveys continuity and disjunction. Punctuating periods serve important purposes: isolating phase transitions, specifying causes, assigning responsibility. Jennifer Wenzel associates this mode of periodization with the geological law of superposition, which assumes that strata closest to the surface are newer. While she questions the implicit hierarchy of linear historiographic models, she also offers a provocative proposal for an Anthropocene boundary event, drawn from leaked Exxon files revealing that the company began covering up climate change research as early as the 1970s, "an intentional act . . . with implications for all life on Earth." "What are the different implications for justice," Wenzel asks, "if one sees history as cyclic or linear, repeated or ruptured, analogous or without precedent?"

As Jeffrey Jerome Cohen observes in his chapter, linearity—with its "definitive beginnings, vexed middles, smoothly inescapable ends"—fails to account for an Anthropocene swirling with "affective detritus, recondite matter, queer fragments, anomalous proximities." Its narrative form is less a sedimented layering, a straightforward plot line, than a tale "sinuous and coiled," what Cohen calls a "vorticular story," "the entwinement of multivectored lines." As an alternative to the linearity of periodization and the search for a definitive boundary event, several of our contributors invoke the geological concept of the unconformity, extending Eric Gidal's pioneering work of Anthropocene literary history, *Ossianic Unconformities: Bardic Poetry in the Industrial Age*.[40] The unconformity is a gap or disjunction in the stratigraphic record that marks a period where no deposits were left or where sediment has been removed by erosion. This break gives form to the intersection of multiple temporalities, forces, or media, just as fossils memorialize a meeting of the biosphere and lithosphere or as ice cores track the history of the atmosphere as coalesced within the cryosphere. Each of these intersections can be understood as an unconformity, where a system has been impeded, disrupted, or enfolded by another and where that disturbance has left a record in formal disjuncture. After all, as Benjamin Morgan insists in his chapter, form is a property of texts and social systems but also of the geological strata and the biological organisms embedded therein.

Extending Gidal's model of bibliostratigraphy, our contributors identify a number of principles by which literary texts establish unconformities insofar as their matter and meaning intersect with broader geohistorical forces, including resonance, precedence, haunting, estrangement, synonymy, anticipation, allegory, cross-hatching, overdetermination, correlation, obsolescence, and coincidence. In some cases, this intersection has to do with the textual medium as matter. Ford points out that in the Anthropocene all writing is "writing on the world" because texts are always haunted by the $CO_2$ emitted in their production. Derek Woods points out that the acceleration of fossil fuel usage in the late twentieth century and into the twenty-first has coincided with a rapidly accelerating production of texts—written, filmed, digital—such that the semiosphere is in effect supplanting the biosphere. Amid this informational onslaught, the Anthropocene also presents us with information loss. Stephanie LeMenager argues that the Anthropocene's casualties include the banality once embodied in the daily news, because "climate change 'news' fails to be 'news' insofar as it implies an end to the everyday itself, since the everyday relies on human habit and its complement of forgetting." Anne-Lise François asks what happens to literature, as a unique medium for the "human bearing of tradition," in an age defined not only by new "technologies of storage and extraction" but also—referring to capital, to the atmosphere's increased absorption of electromagnetic radiation, and to the long-term lithostratigraphic implications of human activity—"by too much retention, too much accumulation, too much permanence."

The unconformity provides a model for reading *absence* itself as a site of meaning, a record or archive. Anthropocene reading often means reading negation, interpreting rifts and lacunae. As Dana Luciano explains in a discussion of nineteenth-century ichnology, fossilized dinosaur footprints indicate "the *presence of an absence*: the mark of the here-no-longer that nevertheless remains." This mark of absence is one of the many ways in which the Anthropocene becomes legible in negation. In his contribution, Matt Hooley points to the fraught implications of narrative absence in relation to Lewis and Maslin's dating of the Anthropocene to the deaths of 50 million Native Americans during the euphemistically dubbed "Columbian exchange" of biota between the Old World and the New. Despite attending to atrocity, Hooley notes, the Orbis hypothesis "makes Indigenous people and knowledge scientifically legible only in or as disappearance." The Native person thus becomes the metonymic figure of vulnerability, the exemplary sign of otherwise diffuse ecological harm. Indigenous disappearance becomes a synecdoche for vulnerability writ large, transforming twenty-first-century Native people into living fossils. LeMenager, too, frames the Anthropocene as a problem of racial invisibility, arguing that climate change eradicates the "privilege of not thinking of oneself as embodied, as matter overwritten and writing history," a privilege accorded by the illusory invisibility of whiteness, which she aligns with the fantasy of living "unencumbered by material constraint."

Unconformities put different phases of the past into contiguity, offering a counterpoint to the linearity often implied by the law of superposition. The unconformity helps us to read instances in which knowledge—partial, anticipatory, or allegorical—of the Anthropocene precedes the term's formal conceptualization. "[W]e hear old things in new ways," as Mentz puts it. Noah Heringman in his contribution tracks the conventions of scientific "romance"—concerned with "negotiating multiple discrepant temporalities"—from Buffon's *Epochs of Nature* (1778) to the popular science of Elizabeth Kolbert's *The Sixth Extinction* (2014) and Jan Zalasiewicz's *Earth After Us*. Taking what she calls a "posthumous perspective," Juliana Chow suggests in her chapter that studies using Thoreau's journals as climate records continue Thoreau's own project, extending the naturalist's work beyond the span of his life while emphasizing its ever-partial, unfinished quality. Ford reads Romantic works of art as "indirectly allusive anticipations" by which the literary artifacts of the past come to write the "Anthropocene present." Nonetheless, he notes, "the Romantics could name themselves but not the Anthropocene," making those anticipatory documents strangely out of step with the era in which they appeared: could not have meant what they now mean. This asymmetrical contemporaneity is the unconformity of the Anthropocene, in which divergent and seemingly incompatible histories rub up against one another, highlighting the potential for the future to remake the past. With this potential for historical unconformity in mind, Justin Neuman, in his contribution, examines how we read climatological forecasts that have not come to pass, turning to Henry Adams's early twentieth-century warnings of global *cooling*, which was to be brought about by the excessive combustion of fossil fuels, a mistaken theory that lives on in the fantasies of climate change deniers even as modernist techno-utopianism recurs in the promise of the geo-engineered "good Anthropocene." Neuman reads Adams's reflections on the technological emporium at the 1900 Exposition Universelle in Paris as an "energy recognition scene," highlighting the way in which recognition may serve as an interruption: the "externalities that extend spatially and temporally beyond a text's representational systems" may intrude upon the act of interpretation.

Reading history in relation to energy flows, which Adams held to be the historian's task in the modern era, complicates not only the linearity of time, but also the idea that it can be divided into units of comparable duration.

The scaling up of human action within the Earth system also entails a simultaneous compression of human history within geologic time. Woods suggests, "As the potential energy of fossil fuels unravels, the speed of history increases" so that "the Great Acceleration is far 'longer' . . . than any other period of literary history" because "there is more history, more communication, and more inscription per unit time than in the past." Cohen characterizes the Anthropocene as "an engine of narrativity powered by acceleration and intensification." It is fitting that Woods and Cohen, who foreground the idea of the Anthropocene as acceleration rather than rupture, study archives far removed from one another—the medieval and the post-1945 "contemporary"—suggesting again how the Anthropocene may be conceptualized as an unconformity within literary history. The Anthropocene proceeds via acceleration and concentration—in François's words, "retention and release," "disinterment and sequestration"—in ways that cannot be neatly periodized.

Periodization, as in the nomination of an Anthropocene epoch, is largely a matter of scale. Events that appear to hinge on dramatic ruptures at one scale become gradual processes of accumulation and acceleration at another. For our contributors, it is the distinct purview of literary genres and forms to make legible such scale variance and translation. In his chapter, Heringman explores the persistence of romance motifs in popular geology. Reading one geologist (Zalasiewicz) reading an influential predecessor (Buffon), Heringman claims that identifying the literariness of geology—its traffic in wonder, its speculative scenes of time travel and truly posthuman reading—serves not to undermine science but to promote a "historical understanding of geological time," the particular ways in which writers adopt literary motifs to convey the scale variances at stake in the Anthropocene. François similarly asks how prose works of Anthropocene science and theory recapitulate the lyric project of "hold[ing] together overlapping yet semiautonomous temporal scales"—"condensing and extending, slowing and accelerating time"—in order to capture the paradoxical temporality of the Anthropocene, a sudden transformation in the scale of human power that depends on the "fast consumption of deep time" and has implications for "the rest of time to come." Morgan approaches the problem of reconciling divergent scales through a renewed attention to literary form, suggesting that "scalar leaps and disjunctures" be approached as *forms*, opening them to "critical strategies for reading mediations, images, and narratives." Morgan argues that Thomas Hardy's novels stage scalar incommensurability, dramatizing our failure to imagine the inhuman immensity of outer space or deep time to the point that "formlessness itself becomes a form." This emphasis on the limits of multiscalar thinking echoes Hooley's emphasis on "nonscalable" ecological vulnerability, an alignment that suggests that rather than continuing to aspire to an encompassing vantage point that would enable us to grasp the full magnitude of the Anthropocene, we should instead make peace with, and even embrace, our inevitable failure to do so. LeMenager returns to the individual scale of the everyday Anthropocene, describing the task of the "Anthropocene novel" as "paying close attention to what it means to live through climate shift, moment by moment, in individual, fragile bodies."

Depending on the date chosen for the Anthropocene's emergence, the identity of the Anthropos changes. Rather than attempting to isolate a single origin story, this proliferation of actors can serve as a guide to the shift from individuals to systems, which is necessary for locating (and addressing) the distinctive causal mechanisms operative in Anthropocene history. In Mentz's chapter, Old Man Anthropos takes the stage to declare, observe, and question his own guilt for an Age of Man that is in fact an Age of Death. This crisis arises from a confrontation with the 1610 Anthropocene, a periodization that attributes responsibility for global environmental change to European imperialism, or, as Hooley calls it, "an ascription" that rewrites "complex, even inscrutable, experiences of environmental harm as readable." In contrast to such resolute assertions of legibility, which enable us to pass judgment, Hooley asks that we "read vulnerably" in and through the impediments to our own understanding, a condition that echoes the practice Wenzel calls "reading under duress." As Wenzel explains, "'duress' derives from *dūritia*, Latin for 'hardness'; it shares this root with 'endure.'" Thus, she asks, "How is reading a form of endurance?" This emphasis on precarious endurance recurs in Cohen's rejection of the "ark" as a bastion against rising seas and climate refugees.

Chow takes a similar approach in advocating for "partial reading," a practice she sees modeled in Thoreau's regional, particular, and perpetually unfinished writings on natural history. In contrast to the systematizing viewpoint pioneered by George Perkins Marsh (often cited as a precursor to Anthropocene discourse), Thoreau's methods offer Chow "a concurrence of biological, literary, and historical forms" predicated on "partialities" and "dispersals" as opposed to "organic wholes" or "monologic continuity." In articulating this vision, Chow adopts a vantage that she calls "critical partiality," which she describes as "a mode of being partial, partial *to* something, partial *of* something." Any act of reading is thus partial in both senses of the term, born of attachment in the midst of incompletion. After all, we cannot read everything. Woods examines this predicament as well. The Great Acceleration that now appears to be the frontrunner for the golden spike also accords with an unprecedented acceleration in media and textual production, threatening to "infowhelm" us at every turn and leaving even the most voracious reader haunted by the "Great Unread." As Woods explains, this overprofusion of texts provides an eerie correlate to the unnamed and unknown species hastening to extinction, further underlining the precarity of Anthropocene reading. In this regard, it seems telling that LeMenager not only highlights the Anthropocene as an occasion for genre innovation in the emergence of "cli-fi," but that in so doing she reaffirms the work of the novel throughout its history: cultivating and expressing the subjective experience of an individual consciousness. The "struggle for genre" is thus also a return, a repurposing, an invitation to read—and read again.

A partiality for literature precedes and underwrites this collection.[41] The Anthropocene is, after all, not what impels us to read. Though our contributors find themselves *rereading* under the sign of this proposed geologic epoch, we were all reading already. We are *partial* readers; sometimes we are too distracted to read or we find that there is just too much to read. Faced with the Great Unread, on one hand, and planetary crisis, on the other, we continue to read as individuals, scholars, and teachers. As Ford observes, the category of literature that emerged in the Romantic period is based on literature's capacity to put "unsayable things"—from absent causality to inarticulate affect—into words. We read because we are terrified. We read to confront our complicity, to ratify our guilt, to mourn the losses. As Heringman reminds us, we read for "wonder," the awe of the Anthropocene sublime mixed with the "pathos" of extinction.

To read is to establish relations in time; reading is revitalization. Luciano describes critical reading as an act of "preservation," a "collaborative or compositional process" arising out of "necessary connections among thought, energy, flesh, mud, minerals, sediment, wind, and water." Reading metabolizes the remnants of absent life into new forms; reading disperses seeds, like the milkweed tufts that Chow traces as they waft from Thoreau's pages into these. To read is to follow Cohen's plunge from familiar archives into history's whirl. In the essays collected here, readings have become writings, interpretations transposed into inscriptions, asking to be read. As Wenzel observes, "allegorizing like the Magistrate, I struggle to read and write in a mode adequate to history, answerable to the future." We read under duress: speaking to the future, about the past, on behalf of the present.

Global geo-engineering projects, resource wars, and refugee crises are all possible features of the near future, but they are not the only ones. Using existing technology, it would be possible to switch the entire world from fossil fuels to renewable energy in a couple of decades.[42] The intractability of the Anthropocene arises from sociopolitical systems rather than geophysical ones. In order to expand our horizons we need to tell different stories, stories that, as LeMenager suggests, help us form attachments "to multiple generations, distant futures as well as distant pasts, all times worthy of curation and song." The Anthropos in the Anthropocene need not refer only to a culpable agent—it can also become an injunction. The species that reads itself in the stone might yet be brought into a new degree of self-awareness *as* a species and, out of that recognition, weave new democracies and inclusive economies, conjoined to resilient ecologies.

At the end of "The Storyteller," Benjamin turns from geology to biology, from bedrock to life: "A proverb, one might say, is a ruin which stands on the site of an old story and in which a moral twines about a gesture like ivy around a wall" (162). From epic forms to proverbs, encompassing wholes transform in slow time—or accelerating time—to fragmentary scraps of wisdom, legible traces of ruination. If the Anthropocene marks a breach in the wall between human and natural history, then imaginative literature may be understood as the ivy that overspreads that wall, finding its way through the gap, entwining the happenings of history, intractable yet fragile, holding onto the crumbling structure even while hastening its decay, wrapping its tangled forms around the ruins of the modern constitution.

## Notes

1. Walter Benjamin, "The Storyteller: Observations on the Works of Nikolai Leskov," in Benjamin, *Selected Writings* (Cambridge: Harvard University Press, 2002), 3:143–66, 147 (hereafter cited parenthetically in text).

2. Jeffrey Jerome Cohen, *Stone: An Ecology of the Inhuman* (Minneapolis: University of Minnesota Press, 2015), 4–6.

3. Bruno Latour, "Agency at the Time of the Anthropocene," *New Literary History* 45.1 (2014): 1–18, 13 (hereafter cited parenthetically in text).

4. See Noah Heringman, *Romantic Rocks, Aesthetic Geology* (Ithaca: Cornell University Press, 2004).

5. Stephen Jay Gould, *Time's Arrow, Time's Cycle: Myth and Metaphor in the Discovery of Geological Time* (Cambridge: Harvard University Press, 1987), 42. Gould emphasizes Burnet's concern with "narrative": "a story line of pasts that determine presents

and presents that constrain futures" (44). In *Principles of Geology* (1830–33), Charles Lyell holds that if history were viewed as proceeding in too short a time span it would "assume the air of a romance," contending that it is only when viewed against the expansive backdrop of deep time that geological change can become legible as the result of slow-moving forces still in operation, an implicit shift from romance to realism. Lyell, *Principles of Geology; or, The Modern Changes of the Earth and Its Inhabitants Considered as Illustrative of Geology*, 10th ed. (London: John Murray, 1867), 94. See discussion in Jesse Oak Taylor, *The Sky of Our Manufacture: The London Fog in British Fiction from Dickens to Woolf* (Charlottesville: University of Virginia Press, 2016), 11–12.

6. Thomas Burnet, *The Sacred Theory of the Earth* (Carbondale: Southern Illinois University Press, 1965), 17.

7. Adelene Buckland argues that for Victorian geologists, the narrative turn was imperative because it served as both "a systematizer of geological knowledge" and a technique for "capturing new audiences and readerships, aligning geology with culturally authoritative narratives from classical and biblical literatures." Buckland, *Novel Science: Fiction and the Invention of Nineteenth-Century Geology* (Chicago: University of Chicago Press, 2013), 17.

8. Will Steffen, Jacques Grinevald, Paul Crutzen, and John McNeill, "The Anthropocene: Conceptual and Historical Perspectives," *Philosophical Transactions of the Royal Society A* 369 (2011): 842–67, 843.

9. Jan Zalasiewicz, *The Earth After Us: What Legacy Will Humans Leave in the Rocks?* (New York: Oxford University Press, 2008), 7 (hereafter cited parenthetically in text).

10. Paul J. Crutzen and Eugene F. Stoermer, "The 'Anthropocene,'" *Global Change Newsletter* 41 (2000): 17–18.

11. Simon L. Lewis and Mark A. Maslin, "Defining the Anthropocene," *Nature* 519 (March 2015): 171–80, 178.

12. See Bruno Latour, *We Have Never Been Modern*, trans. Catherine Porter (Cambridge, Mass.: Harvard University Press, 1993).

13. Dipesh Chakrabarty, "The Climate of History: Four Theses," *Critical Inquiry* 35.2 (2009): 197–222.

14. Jason Moore, *Capitalism in the Web of Life: Ecology and the Accumulation of Capital* (London: Verso, 2015), 1–2.

15. Immanuel Wallerstein, *The Modern World-System*, vol. 1, *Capitalist Agriculture and the Origins of the European World-Economy in the Sixteenth Century* (New York: Academic, 1974), cited in Lewis and Maslin, "Defining," 175.

16. Martin Rudwick, *Bursting the Limits of Time: The Reconstruction of Geohistory in the Age of Revolution* (Chicago: University of Chicago Press, 2005), 162–63.

17. See discussion in Lewis and Maslin, "Defining," 177.

18. Francis Fukuyama, "The End of History?," *National Interest* (Summer 1989): 3–18; Bill McKibben, *The End of Nature* (New York: Anchor/Doubleday, 1989). On this conjunction, see Margaret Ronda, "Mourning and Melancholia in the Anthropocene," *Post*45, June 10, 2013, http://post45.research.yale.edu/2013/06/mourning-and-melancholia-in-the-anthropocene.

19. Colin N. Waters et al., "The Anthropocene Is Functionally and Stratigraphically Distinct from the Holocene," *Science* 351.6269 (2016): aad26221–10 (hereafter cited parenthetically in text).

20. See Johan Rockström et al., "Planetary Boundaries: Exploring the Safe Operating Space for Humanity," *Ecology and Society* 14.2 (2009), https://www.ecologyandsociety.org/vol14/iss2/art32.

21. Jan Zalasiewicz et al., "Colonization of the Americas, 'Little Ice Age' Climate, and Bomb-Produced Carbon: Their Role in Defining the Anthropocene," *Anthropocene Review* 2.2 (2015): 117–27. The authors also cite Lewis and Maslin, who consider both a 1610 start date and "a peak of radioactivity" in 1964, associated with nuclear weapons testing. Lewis and Maslin observe that the disadvantage of the 1964 radiocarbon dating is that "although nuclear explosions have the capacity to fundamentally transform many aspects of Earth's functioning, so far they have not done so, making the radionuclide

spike a good GSSP [global boundary stratotype section and point] marker but not an Earth-changing event" ("Defining," 177).

22. Clive Hamilton, "Getting the Anthropocene So Wrong," *Anthropocene Review* 2.2 (2015): 1–7, 5. "Finding new species (or other signs) in rock strata," he observes, "is not the same as identifying a change in the functioning of the Earth System" (5).

23. Jacques Derrida, "Signature Event Context," in his *Margins of Philosophy*, trans. Alan Bass (Brighton, England: Harvester, 1982), 307–30, 328.

24. Zalasiewicz et al., "Colonization of the Americas," 122.

25. As Rudwick shows in *Bursting the Limits of Time*, in the later eighteenth and early nineteenth centuries the earth sciences were divided between classificatory and speculative imperatives. Geognosy, closely associated with mining industries, is a science of description and classification, not of "causal explanations" (84). Geognosy gave way to what William Smith calls "stratigraphical geology," which uses fossil remains to delineate formations. Rudwick contrasts this approach, which establishes its scientific credentials on the modesty of its empiricism, with the work of Georges Cuvier, a French paleontologist whose observation of significant extinction events invited the speculative reconstruction of the forces of past geologic cataclysms.

26. Andreas Malm, *Fossil Capital: The Rise of Steam Power and the Roots of Global Warming* (London: Verso, 2016), 390–91.

27. Haraway coined the term in conversation with fellow anthropologists Noboru Ishikawa, Scott F. Gilbert, Kenneth Olwig, Anna Tsing, and Nils Bubandt. See their "Anthropologists Are Talking—About the Anthropocene," *Ethnos* 81.3 (2016): 535–64, 557.

28. Bronislaw Szerszynski, "The End of the End of Nature: The Anthropocene and the Fate of the Human," *Oxford Literary Review* 34.2 (2012): 165–84, 180.

29. Will Steffen et al., "Stratigraphic and Earth System Approaches to Defining the Anthropocene," *Earth's Future* 4 (2016): 324–45, 337.

30. Fredric Jameson, *The Political Unconscious: Narrative as a Socially Symbolic Act* (Ithaca: Cornell University Press, 1981), 56.

31. See, for example, Rita Felski, *The Limits of Critique* (Chicago: University of Chicago Press, 2015).

32. Stephen Best and Sharon Marcus, "Surface Reading: An Introduction," *Representations* 108.1 (2009): 1–21.

33. In relation to the Anthropocene, some of the most productive work in digital humanities may be that attending to the ecological embeddedness of media technologies themselves, suggesting an alignment with fields such as book history as routes into the materiality of the text. Examples include Jussi Parikka, *A Geology of Media* (Minneapolis: University of Minnesota Press, 2015); and Nicole Starosielski, *The Undersea Network* (Durham: Duke University Press, 2015).

34. Caroline Levine, *Forms: Whole, Rhythm, Hierarchy, Network* (Princeton: Princeton University Press, 2015).

35. See, for example, Julian Yates, *Of Sheep, Oranges, and Yeast: A Multispecies Impression* (Minneapolis: University of Minnesota Press, 2017).

36. See Wai Chee Dimock, *Through Other Continents: American Literature Across Deep Time* (Princeton: Princeton University Press, 2006); David Damrosch, *What Is World Literature?* (Princeton: Princeton University Press, 2003); Franco Moretti, "Conjectures on World Literature," *New Left Review* 1 (January–February 2000): 54–68.

37. Rob Nixon, *Slow Violence and the Environmentalism of the Poor* (Cambridge: Harvard University Press, 2011); Stephanie LeMenager, *Living Oil: Petroleum Culture in the American Century* (Oxford: Oxford University Press, 2014). Early ecocriticism's resistance to the linguistic turn is evident in many of the essays collected in Cheryll Glotfelty and Harold Fromm, eds., *The Ecocriticism Reader: Landmarks in Literary Ecology* (Athens: University of Georgia Press, 1996).

38. See Jeremy Davies, *The Birth of the Anthropocene* (Berkeley: University of California Press, 2016), 140.

39. Ted Underwood has identified a similar predicament in literary history, noting

that periodizing terms like "Romanticism" or "modernism" are predicated on the ability to read stark breaks located around specific events or in the work of specific authors. "In principle," Underwood argues, "literary scholars should be able to move back and forth between different kinds of historical argument, invoking continuity or contrast as necessary for a particular thesis. But in practice, we find it very difficult to make arguments about continuous, gradual change." *Why Literary Periods Mattered: Historical Contrast and the Prestige of Literary Studies* (Stanford: Stanford University Press, 2013), 169.

40. Gidal in *Ossianic Unconformities* adopts the Huttonian principle of the angular unconformity, "disjunctions in the stratigraphic record, ... physical manifestations of heterogeneous time," as a way to read similar compressions and dislocations in the "poetic unconformity" of the Ossian poems, such as the interweaving of elegiac and progressive moods, and in the "medial unconformities" of Ossian's nineteenth-century reception as the poems were remediated in ordnance surveys, musical scores, and tourist guides (5, 7, 68). Gidal identifies in Ossian's reception a persistent cross-referencing between text and topography. This "biblio-stratigraphy," as he terms it, provides an acute mode of reading "the social and spatial disruptions of industrial modernity" (15, 12). *Ossianic Unconformities: Bardic Poetry in the Industrial Age* (Charlottesville: University of Virginia Press, 2015).

41. In making these claims for reading, we are not attempting to claim the Anthropocene as the exclusive province of literature. After all, Benjamin balanced his investment in story with a fascination for film, celebrating the camera's capacity to render phenomena "accessible only to the lens (which is adjustable and can easily change viewpoint) but not to the human eye" through effects like "enlargement or slow motion, to record images which escape natural optics altogether." Imperceptible phenomena, such as climate change, become knowable only through an expansive array of media—from computer models to ice cores, from films to video games—and hence all Anthropocene readings are multimodal. Benjamin, "The Work of Art in an Age of Its Technological Reproducibility: Second Version," trans. Edmund Jephcott and Harry Zohn, in Benjamin, *The Work of Art in the Age of Its Technological Reproducibility and Other Writings on Media*, ed. Michael W. Jennings, Brigid Doherty, and Thomas Y. Levine (Cambridge, Mass.: Harvard University Press, 2008), 19–55, 21.

42. Malm, *Fossil Capital*, 368–69.

# 1.
# Anarky

*Jeffrey Jerome Cohen*

*And why should I not drown with friends? Let the waters wash. Ale and song and a love of this convicted Earth exceed that gated menagerie, stockpile for a narrowed future. If the floods must scour, if the mountains sink, then think of me when you find an ammonite on the new world's hill. Place it to your ear. Listen to your heartbeat sea, its pulse and punctuation. Know that abyss and vault roared a deluge to inscribe a story without sufficient voice. We sang a round and drank from bottles. Water puddled and then swelled. When gate and gangway closed, they left my companions to the weather.*

•

Human history is shallow and local, a chronicle of small wonders, rapid in its tempo. Geologic time is cosmic and profound, a dense account of planetary formation, mineral thriving, continents that slowly glide on a liquid mantle, ceaseless subduction. Attenuated tale of a single species, human history is a segment cut from an overwhelming story. Deep time's scale is a mathematical line that extends to infinity. A forty-day deluge cannot punctuate a record inscribed by zircon, snowball Earth, fossilized monsters, a story puzzled from recalcitrant stone. Human history is conveyed by a storehouse of translatable texts and loquacious objects, lush in possibilities. Interpretive practices derived from a linguistic archive will therefore not be of much use when reading this geological repository.

Yet as Noah Heringman has argued, the era we want to christen the Anthropocene marks an unprecedented "act of writing ourselves into the rock record," a new chapter in planetary history characterized by "radical discontinuity" with preceding pasts, human and geological.[1] During the Anthropocene, human time and geological time find their confluence, enmeshment through lithic inscription.[2] Two seemingly incommensurable temporalities and their modes of reading flow together, propelled by force and trace and an ardor for leaving pasts behind. *And there a period.* This recent union, however, also makes evident what Heringman calls "a kind of textual materiality in geological events" ("Deep Time," 58), something like human writing present in the lithic record all along. To sentences and to epochs belong punctuation marks, formed by pen or sediment, pixel or fossil.[3] To narratives and to eons belong quiet plots, struggles against extinction or death, sedimented archives of story.

Whether deep, close, surface, or distant, contemporary reading strategies love the linear: definitive beginnings, vexed middles, smoothly inescapable ends. Through lines and strata, the progress of narrative and time becomes legible. This chapter's project, however, is to drift some eddies instead, a vorticular topology of reading.[4] A vortex is that dynamic form that spins into being at the meeting of heterogeneous pressures and trajectories, admixture in swirled motion. At any given moment Earth's atmosphere has at least one vortex spinning, churning horizontal layers into multiplanar trajectories. A vortex often obliterates. Yet its spirals and shifts may also render evident long-standing embroilments, the unexpected touchings of forces and entities within expansive spatial and temporal systems. Vortices are disjunct histories in contiguity, binding lines into curved motion, a model of temporality that does not easily sediment into discrete layers. Most Anthropocene narratives embrace linearity: periods possessed of hard starts and full stops, plots with rising action, accelerated propulsions, catastrophic denouements. Yet their currents swirl with affective detritus, recondite matter, queer fragments, anomalous proximities.

Most writing on the Anthropocene grapples with the problem of determining its dawn: with the advent of widespread agriculture and resultant deforestation? With the possible carbon maximum of 1610 and Europeans' violent movement into the Americas? With the patenting of the steam engine? With the Great Acceleration? With the embedding of soot, plastic, concrete, radiation from atomic bombs, or the detritus of factory farming in the rock record? A question that might be worth the lingering, however, is whether the Anthropocene is a segment cut into Earth's arrowlike timeline, an era initiated at a specific and determinable date, or a more diffuse designation for "forms of time being transmitted by . . . other-than-human media . . . expanding, counterpointing, and complicating human conceptions of historical time, not overwhelming them," the topography of which might be more sinuous and coiled.[5] Given its geologic scale and force, need the Anthropocene hinge upon narratives straightforward and discrete? Might it not perturb familiar chronologies and convolute customary modes of sorting, apprehending, and narrating human time? What forms might emerge for reading the Anthropocene if we do not anchor its *arche* ("origin," "rule," "authorization") in some trace of carbon or radioactivity, some precise percentage of species loss or ocean acidification or deforestation? What might the Anthropocene become if it is not tethered to interruptive punctuation and delineative origin?

Mick Smith has articulated what he calls an anarchic ecological ethics, "not dependent on defining in any absolute or authoritative form its worldly origins or task."[6] Tentative and mundane, aware of its own fallibility, attentive to "the flows and depths of diverse worldly existences happening beneath their surface appearances" (*Against*, 64), Smith's praxis reads not for depth but for volution within what only seems to be stability. What if our attention were not contained by what happens to have been preserved in the ark (*arcus* is the Latin word for a chest or archive) but saw what eddies under, around, and sometimes through that repository, carries it back and forth, time itself not as laminar flow but as spiral of unforeseen propinquity? Stories chosen for preservation speak loudest. They bestow an enduring but flattened shape on history—and that form comes to seem, through long tradition, an inevitable plot line. Yet much remains and even flourishes within the billows and whorls surrounding such archives, offering tales more sinuous and coiled.[7]

•

I'll come clean from the start. I do not believe in periods or genres, not really, not with sufficient enthusiasm. Certain gatherings may compose orderly shapes from a mess, organize heaps and muddles into tidy chests. Yet on closer examination an imbroglio remains. The outside clamors for admittance. Forms effloresce, and culture-bound modes of story live for a while, flourish, catalyze new narratives, go dormant or extinct. But not everything can be read back from denouement. Periods are partial stories, and stories do not end, punctuate as we may. Plot lines thrive and mutate beyond artful terminus, beyond historical probability. Comedy is tragedy is history is romance is comedy. We dwell within a contemporaneity that encompasses the steam engine and the atom bomb. Our present also includes the global dissemination of agriculture and its attendant tumults as well as "the catastrophe (if you are anaerobic) called oxygen."[8] Every expansion of historical frame entails a compression of events. So many persisting conditions affect the possibility of clearing out a *now* to make our own that the present is best described as what Timothy Morton calls "an ever-widening set of concentric temporalities" ("How," 264). No wonder I sometimes have trouble telling the difference between a romance composed in medieval French lamenting the conclusion of the Arthurian age and a critical essay replete with Anthropocene melancholy—or (at my worst) between lyrical ekphrasis and a satellite image of a storm that spins against our dwindling coasts.

Poiesis is a making. I have learned from Roger Caillois, Marie de France, Jane Bennett, and Manuel DeLanda (among many other fine posthumanists) not to worry too much about allocating composition to singular entities. Medieval writers told tales of gems with lyrical powers not because they were naive but because they perceived within lithic density the force of a world not fully theirs, vertiginously productive, a confluence of the human and the geological, a companionship in story making and a convolution of durations that it has taken us until the Anthropocene to recall. The framing of this new epoch conveys stories long intimate to human history, cataclysmic narratives in which human action and inhuman force ally. "Cataclysm" derives from the Greek word for the biblical deluge, *kataklusmos*, a frame for a history punctuated by disaster.

Periodization is an act of sovereignty: lines out of whorls, stories from worlds. Those who resist such delineations risk losing their narratives, their lives. In some medieval texts, Noah's wife declines to board the family's houseboat, rebuffs a dwindled future. In the Chester play *Noah's Flood*, she attempts to remain with her companions until forcibly taken aboard. The medieval drama is a public restaging of the deluge, from Genesis, mounted as part of a cycle of performances that carry biblical history into a communal present. Along with the York and Towneley dramatizations of the flood narrative, the Chester play is notable for the disruptive force exerted by Noah's unnamed wife, who refuses to depart a world she is not supposed to love.

•

*Embody your countercurrents in an angry woman—a wife, no less—and let her rail against the flood. But then a son will descend to force her aboard while friends remain in the deepening sea, wielding carols and ale against the surge. "The fludd comes fleetinge in full faste."*[9] *Our ark was built against a fallen world, its membership exclusive, walls raised against climate refugees. History is written with force. "One everye syde that spredeth full farre." I wanted to remain in the storm so that another story, sung as a round, might endure. "For fere of drowninge I am agast." The waters poured from the vault of heaven, and the waters gushed from the abyss. You will have forgotten that the flood was confluence, its shape a gyre. "Good gossippe, lett us drawe nere." When all the Earth was ocean, we could not tell the valleys. Everything smoothed, and our ship became a singular story. But sometimes below the water I glimpsed the drowned. Their song was held aloft, long after, its archive restless air and water. History is read with force.*

•

The narrative repertoire we humans possess is splendid in possibility but limited in matter, pattern, and plot. We reuse stories and create palimpsests.[10] We compose through alliance. The creative and archival repertoire that stone possesses is likewise splendid in possibility but limited in material, pattern, plot. We may have learned to intensify our art from this enduring substance, the substance upon which all biological flourishing depends (organic life is geochemical).

The Anthropocene as the great rejoining of two partners in world making is the belated recognition of an abiding tale, the opening of an obscured archive, a challenge to the topology of customary reading practices. With delimiting sides fashioned of varied materials (wood, stone, ice, air), every archive is a box through the agency of which an inadequate number of records are preserved, a climate-controlled collation secured against time and weather, *le temps*. As anyone who speaks French knows, or who sees in the word "weather" a verb as well as a noun, history and climate are the same thing. Michel Serres observes, "Time is paradoxical; it folds or twists; it is as various as the dance of flames in a brazier—here interrupted, there vertical, mobile, and unexpected. The French language in its wisdom uses the same word for weather and time, *le temps*. At a profound level they are the same thing. . . . Time flows in a turbulent and chaotic manner."[11] Time/weather (*le temps*) is vorticular rather than smooth. Its elements are swift air, fire, and water—and earth within its proper duration (stone is slow water). Ian Baucom argues that the Anthropocene demands new topologies of time, emphasizing a "temporal and ontological multiplicity" that knots the progressive lines of customary history.[12] I am sympathetic to Baucom's call for an Anthropocene-engendered "fourth order of history, measured both in dates and in degrees, in times and temperatures; an historical, infra-historical, and supra-historical order" (142)—but with Serres I believe that Baucom's corkscrewing version of time/climate has been with us all along. We do not need to grant the Anthropocene its desired ruptures.

We treasure what we have with long and serene habit conserved against *le temps*, so that a curated collection of records that might have originated in happenstance sediments into historical strata, the straight lines of reading.

Yet sediment is "earthy or detrital matter deposited by aqueous agency."[13] Its accumulation is a long record of wash and flood. We can learn as much from what has been left to the tempest ("time storm") around the ἀρχή (arche), the building or room or chest into which inscriptions of a partial story have been placed to buttress hierarchy, authority, and other straightened or stratified slices of history. Yet our long apprenticeship in the knowledge of how to read the strata credentials us as disciplinary experts, literary geologists who specialize in one of the archive's time stripes.

•

Stratigraphy peers into lines (strata) of stone to discern the inscriptions (graphesis) of long narratives, Earth's history made legible through layering. That this close reading strives for the linear, with each successive striation well punctuated, is underscored by the idea of the "unconformity," a discontinuity in the geological record where horizontal bands have shifted to sit at unforeseen angles. When James Hutton observed such strange rotations he was provoked to formulate an aeonic approach to geology that came to be known as uniformitarianism.[14] Emphasizing the steady unfolding of repeating, incremental processes, uniformitarianism is a patient hermeneutic that contrasts with catastrophism, the view that the geological record is a story of havoc, a register of sudden floods and fiery irruptions. Through what he described as "reading in the face of rocks the annals of a former world," Hutton assisted mightily in the liberation of archive Earth from fleeting calamities and the mayfly meter of human time.[15] World-engulfing disasters like the water against which Noah built his ark cease to matter, except perhaps as unremarkable manifestations of endlessly recurring planetary processes. Catastrophe narratives are brusque, anthropocentric, and not the stuff of stones. In the expansive scales that Hutton's lithic annals demand, "we find no vestige of a beginning—no prospect of an end."[16]

Yet Hutton's reading for depth depends on unsettled texts, petronarratives that defy their contexts. His inspiration derived from the cliffs of his native Scotland, a geological repository in which he beheld what his companion John Playfair described as "the abyss of time," a story writ in stone that unfolds "too slow[ly] to be immediately perceived . . . the duration of the world."[17] Hutton's epiphany arrived when he realized through a combination of close reading and imagining the massive distribution of his data that restless rocks refuse any plot that could unfold at human pace. The garrulous angles of geological unconformity are the bucking of terrestrial force against anthropocentric intelligibility. Such convolutions in the archive can be narrativized only when the

Earth is considered within its proper duration. Deep time moves slowly forward, with breaks that inscribe transitions between eons, eras, periods, epochs, and ages—and yet in its roiled flow will also reveal a tendency to curve back upon itself, engendering whorls.

Unconformity triggers the imagination, an invitation to creativity and inventive apprehension, to the realization that the lithosphere flows not like some tranquil river but like the billowing sea. Waves of strata crash so slowly that we discern no movement, yet their surge and batter are more powerful than any ocean ever voyaged.

•

*When they compelled me to load the boat, I ushered to the hold untreasured things. They love their lions and their sheep, their animals for allegory and aliment. I am the guardian of derided beasts, the chaotic and the fierce: burglars of human plenty, creatures of dissent and scorn. Not every animal is housebroken, householded, housemaking. Not every skin is vellum or sacrifice. Bears, wolves, weasels, apes, owls, squirrels: I would proliferate them all, and so I stowed them in snug pens.*[18] *Does this family think a future forms without dissension, without triggers to story? They would have peace in a wave-tossed ship. I do not think a wooden wall can ever hold against such tumult. The ocean is already within. Like the air, the characters it carries endure far longer than men and gods declare. Though built against the drowned, more stories cling to the ark than you have been taught.*

•

The International Commission on Stratigraphy publishes a chronostratigraphic chart that maps geological progression from the extensive Hadean to the brief Holocene, the epoch in which—at least for the time being—we officially dwell.[19] This ten-thousand-year span includes the retreat of the glaciers, whose stories are inscribed in everyday stones, the proliferation of forest-clearing agriculture, the extinction of the mastodon (along with countless other species that were our contemporaries and companions), genocide, worldwide petrochemical addiction, and a human population explosion. The ICS is slowly weighing the possibility of declaring that we inhabit a new epoch, one even more recent than the Holocene ("wholly new"). Starting with widespread farming, or with a planetary dip in $CO_2$, or maybe with the patenting of the steam engine, or perhaps with the Great Acceleration, or most likely with the blasting of atomic bombs, we may have initiated the Anthropocene, the Age of Man, the inscription of a punctuation mark after which all alters.[20] With its ardor for disequilibrium,

Anarky

storm, speed, technology run amok, and sudden havoc, the Anthropocene is a catastrophist's dream. Yet stratigraphy is a science of straight lines and sedimented narratives, a story built on layers. Its love of straightforward sequence is so forceful that the unconformities within the Anthropocene's rupture-obsessed narrative are in danger of being smashed and leveled before they speak a more vorticular story. *And there a period.*

The word "period" derives from the classical Greek *peri-* ("around") + *hodos* ("going"), a circling more than a punctuated, linear segment. Medieval Latin employed the noun *periodus* to denote a cycle that recurs, like medieval performances of biblical narratives—and "recur" itself means "to run back, to circle 'round." Linguistic strata subjected to close reading reveal themselves full of signifying whorls, at least according to the geology of language, etymology. But if the Anthropocene is the new wholly new that comes after the wholly new, familiar terms and reading practices may only obscure the apprehension of a transformed reality, a narrative without familiar parameters. Perhaps the Anthropocene does not simply initiate an epoch (if only we could determine at what precise date it commenced) ignited by rupture, so much as coil story and time, an engine of narrativity powered by acceleration and intensification.

•

*So we stood, probably on a rock, but after all this time who knows? The waves were rising. We were giddy women, and we were drunk. "The fludd comes fleetinge in full faste."*[21] *We sang our round against the tempest. Our ankles were wet, then knees, hips, chests. "And lett us drink or wee departe." We boozed. We crooned. When the water reached our necks, we embraced and tasted salt. "Good gossippe, lett us drawe nere." At last from the gangplank, my sons descended, first to plead and then to force.*

> Japhett: Come into the shippe for feare of the wedder.
> Noes Wyffe: That will I not for all your call but I have my gosseppes all.
> Shem: In fayth, mother, yett thou shall, whether thou will or nought.[22]

*From the little window I watched my companions cling to their rock, their bottle, their song. By the time the ark lifted they were gone.*

*So departed the world. I knew that when the waters receded and the ship came to rest on a mountain we would sacrifice some cherished animals. The clean beasts and their children we would eat, use their bodies for tools and for art. You will have forgotten that before the flood, humans feasted only on plants. The animals we immolated or devoured when we left the ark had been our messmates, our companions.*

Like stratigraphy, the reading practice called historicism holds that the literary record is linear, well punctuated, the product and legible bearer of roiled climate stilled into context. Texts and genres are signifiers of circumstance, best interpreted by looking closely at economic and social conditions, predecessor texts and contemporary events, cultural contact, and social and intellectual transformations, all of which exert determinative force. Stratigraphy has epochs; historicism has periods. The literary archive is horizontal, layered, punctuated—even if we have more difficulty than we admit discerning the integrity of its strata. We cannot expect to find the matter of medieval romance within a nineteenth-century lyric unless a writer is deliberately old-fashioned, nostalgic, or anachronistic (and even then, that impulse is mainly legible as a flight from contemporaneity rather than as the conveyance of temporal anomaly). Yet like the geological record, literary texts and the archives to which they belong bear unconformities.

Eric Gidal defines textual unconformities as "more poetic" versions of the spiraled strata that catalyze geology's reading practices: untimely irruptions within a narrative, confounding attempts to render a work the record of a moment.[23] A textual unconformity resists readability, preventing the flow of a smooth "historical continuum between past, present, and future" (*Ossianic Unconformities*, 183). Such a lively temporal admixture challenges "stadial history," the presumption that poetics as well as peoples progress into better forms over time (with "better" meaning "more like us," assuming ourselves an apex). Gidal's unconformities are convolutions of time readable in both the literary and geological records—or, better, they are geotextual whorls that render stratigraphy and landscape history inseparable from the interpretation of the texts upon which they are thought to have left a palpable imprint. Gidal describes such unconformities as "uniquely modern," since they convey remnants of histories lost when landscapes metamorphose under the pressures of industrialization. No coincidence therefore that the Ossian poems Gidal studies, the invention of the steam engine, and the formulation of geological deep time are roughly contemporaneous. The Anthropocene thereby marks the human race becoming an unconformity, self-immortalization as a "trace preserved forever in the rock" to disrupt future reading practices (184).

Little of the actual past is to be found within Gidal's unconformities. The segment they cut from time is a small one that moves quickly forward. Yet James Macpherson's Ossian poems are set in medieval Britain, in a period during which the archipelagic kingdom of Dál Riata enabled a constant flow of people, languages, goods, and stories to traverse the coasts. Dál Riata vanished over time into an anglicized Britain, but that assimilation required long, violent work. In the wake of the Norman Conquest, England was for a time annexed to a transmarine empire stretching from northern Europe to the Levant. We speak with such reverence of an age of discovery in the early modern period that we forget that before Europe encountered its New World it had to discover itself. Much of this discovery unfolded through the "vessels of modernity": ships of trade, exploration, and war.[24] As various depictions of Noah's ark suggest (chest, long ship, cathedral of the seas, cog, or carrack), ships are always untimely, always modernity incarnate, and yet "vehicles of mythic consciousness."[25] Water does not periodize like stone or landlocked texts.[26] Its archive eddies, whirls, conveys dangerously, transforms the submerged into the rich and strange. Noah knew this in his ark, its wooden walls preserving some creatures from the sight of those who drowned. Through tempests, shipwrecks are perpetual, but so are intimacies and rounds.[27]

Europe had to be made, a ceaseless process of communal fabrication and internal assimilation with no guarantee of success. A fragile collective, this shifting amalgam was deeply heterogeneous, never inevitable. Numerous indigenous peoples suffered or were obliterated; heterogeneous cultures, languages, and narratives were assimilated or destroyed so that the European nations we know today could become themselves. Richly vernacular stories were left outside the archive. Much was lost to the weather. National homogenization, interior colonization, the forceful imposition of regularity, and genocidal war are easy to forget if we imagine that those galleys traversing the Atlantic set sail from entities forever stable. The Americas at which they arrived were deep in an Anthropocene of their own, their terrains having been massively reconfigured through a variety of land management practices. The 1610 date for the Anthropocene has the unintended consequence of once again obliterating Native American histories, of telling a familiar tale of European discovery that features indigenous peoples as the already disappeared.[28] History is contingent, filled with possibilities or unconformities straightened into stadial accounts only through great force.[29]

Gidal's sonorous term "biblio-stratigraphy" (designating the accumulation of commentary on a lively work) could also denote the thriving of any popular medieval text as it moves from one vellum instantiation to another, birthing all kinds of pasts and futures, exegeses and sequels. Manuscript culture is of necessity bibliostratigraphic since most recensions of a narrative will be imprinted by multiple temporalities, knots of disjunct inscription. *Beowulf* conveys stories dating from at least circa 516. Though the Old English poem may have been composed as early as the eighth century, the manuscript we possess was inscribed circa 1000 C.E., and we cannot know how many copies preceded this version.[30] Macrobius's fifth-century *Commentary on the Dream of Scipio*, which records an early impetus to imagine the beauty of the round Earth as viewed from the cosmos, inhabits the philosophy of Boethius, the Arthurian romances of Chrétien de Troyes, and the cosmological imaginary of Chaucer's *Parliament of Fowls*. Scipio's dream of space travel was first narrated in Cicero's *De re publica*, lost for millennia before its 1819 rediscovery in the Vatican library, where it was hiding in plain sight. The vellum on which the long-sought text had been inscribed was at some point scraped and reused, so that at first glance the skin bore only a work by Augustine of Hippo. A manuscript that holds such a simultaneity of impressed stories is called a palimpsest. Were we being more geological, we might label such objects "vellum unconformities."

•

How would a mode of reading that abandons the stratified, punctuated archive proceed? Not one text layered upon another, but time convoluted into a whorl. Literary analysis need not replicate stratigraphy, even if familiar reading practices share tools and assumptions with geology. Lines are cut from spirals, just as spirals are the entwinement of multivectored lines. We need not assume an absolute break between deep time and human history, between rocks and water, science and art, periods and epochs, the Middle Ages and the Anthropocene. Some suggestive possibilities for gyred reading are found in the long history of Noah and his ark, a narrative machine that does not stop generating alternatives to inundation's sufferance. Alongside this foundational story of catastrophe, climate change, and gated community, dissenting tales have perennially unfolded that embrace the perspective of the excluded, that posit a transhistoricism of affect in an attempt to feel across unbreachable walls. The root of the English word "ark" is found in the Latin *arca*: a chest for safekeeping, a source of political authority, a point of origin, the arche of archive. A place of immurement and preservation, an archive is selective. Much must remain outside. What if Noah's arkive (a floating strongbox of human and animal stories, of trans-species possibilities in the face of ecocatastrophe) is abandoned as the singular origin?

The long tradition of dissonant stories that unfold outside the parameters of the ark is well conveyed by the Chester play *Noah's Flood*. Deluge stories impede easy periodization: they look back to a past that includes the Levantine flood stories familiar from texts like the *Epic of Gilgamesh* and forward to the frequent framing of contemporary climate change as a Genesis-style flood narrative. They also resist reduction into totalizing, Anthropocene-style tales about the human as an abstract, disembodied force. The Chester *Flood* drama gives voice to the earthbound desires of Noah's wife, who chooses her drinking companions over immurement in the family ship. She calls the ark a "chiste" (chest) while clinging to her "pottle" of shared booze. Her willingness to remain in catastrophe and community rather than imagine a salvific moment of rising above the waters is a resistance to linear history, a curving of its trajectories against its own love of rupture. Epochal narratives attempt, sovereignlike, to periodize. Yet as Noah's wife insists, these stories are corporeally, affectively, violently, and unevenly felt. They do not easily transcend embodied particularities. When the Chester play was performed, Noah's wife would have stood with the audience, who were likewise excluded from the ark, standing with the drowned.

Although taken to be an exemplary work of medieval drama, the best version we possess of the Chester *Noah's Flood*, the version haunting this chapter, dates from the early seventeenth century. Its contemporaries are texts like Shakespeare's drama of shipwreck and perseverance *Pericles* and a boom in pamphlet-based disaster porn. The story of the ark seldom settles into an era within which it can be separated off into a fullness of contextual meaning, arriving instead as a perpetual unconformity, not amenable to clean periodization, easy climatic impress, or the stratigraphic archive. The ark should not float and yet it does, a vessel built for maelstrom.

•

Close and slow in its reading, historicism totalizes. Enamored of discontinuity, it aims to demonstrate why (for example) Shakespeare's characters express a subjectivity, individuality, market savvy, global consciousness, sexuality, free will, agonistic relation to a deity, and intimation of the Anthropocene that preceding centuries could not. Emergence will often attenuate rupture, but the point where difference becomes remarkable always seems consonant with the period in which the literary interpreter has been trained and finds an archive (even as the period during which the interpreter lives also impresses itself upon the questions asked and discoveries made: different pasts arrive from different presents). Differences in intensity become sudden differences in kind. The hegemony of periodization meanwhile remains intractable. Reliant on what Julie Orlemanski has called a "fragile logic of exemplarity" that sutures together "reading and history,"[31] historicism divides the literary record into marbled strata, characterizes each through what has become newly possible, and reads in texts the traces of that climate change. In any archival trace may be found the world at its moment of composition. If the Anthropocene began in 1610, then *The Tempest* will be replete with its legible signs. The indigenous peoples of the Americas take a final gasp of air as Caliban curses his enslaving master. The storms swirled by Prospero through his rough magic portend the hurricanes that batter American coasts.

The confluence of Lewis and Maslin's early modern golden spike, Shakespeare's late play, and the burgeoning of the Anthropocene makes legible a knot in which present and past are confoundingly entwined.[32] But even such entanglements do not necessarily disrupt the topography of totality-obsessed historicist reading or the linear temporalities upon which literary stratigraphy depends for its narratives. Presence is not causality, and history holds more intensification than advent. Orlemanski writes that the "synecdochic logic" ("Scales," 223) of exemplary reading relies on a totality that must be discernible from a distance—just like the Anthropocene, the era that we can imagine only from what Jan Zalasiewicz and colleagues call "the perspective of the far future."[33] This retrospective coherence would include all differences and form a set or system, segmented into a period. Yet if a totality is actually to encompass all differences it will never stop circling back upon itself. The period as segment or line connects in motion past, present, future—and becomes thereby the round, the cycle, the spiral. Time does not flow so much as coil.

•

The Anthropocene marks a rupture, gathering the Earth together at last with humans, climate, oceans, stones, animals, soil, ice, atmosphere, forests, shorelines, desert, fungi. Designating an epoch in which none of these are undisturbed, the Anthropocene offers a focus point. It assists in thinking through the profundity of contemporary ecological challenges. Scale zooms out vertiginously, from atomic stories in lithic strata to a planetary whole, from primordial stone to far-flung futures, from close reading to distant processing. But difficulties attend such swift scalar shifts. The Anthropocene offers such a distant reading on past and present that specific and meaningful differences, human and historical, easily vanish. So do unconformities, which flourish within difficult middles rather than at the extremes (and all middles—including the Middle Ages—can only be made middle to other events; they are not naturally aligned within progress narratives).

The Anthropocene promises to name a problem and thereby ameliorate our anthropogenic nightmares, if only we can let go of the mundane differences that will enable the Anthropocene totality to close and a detached or distant perspective to be achieved. But why not admit the Anthropocene is a failed encompassing: a heterogeneous yet vibrant tempestuousness that as it attempts completion of its circuit finds its course altered? There's no return to arche or origin that is not a spiraling. What if the Anthropocene names not a period, which relies on linear, stadial, punctual notions of temporality, but a whorl?

•

We have zoomed back from the Earth to behold the globe suspended in space, a radiant sphere, the famous blue marble (seen in an Apollo 17 photograph taken from forty-five thousand kilometers, 1972). We have pulled so far back that we have beheld our planet as a pale blue dot (seen in a *Voyager* photograph taken from six billion kilometers, 1990). Both images offer an icon of the Anthropocene, a "uniquely modern" moment of humans viewing from a distance the only home they have known. It seems we have to leave the Earth to know it. The blue marble and the pale blue dot diminish human self-importance—even as we ruin the very planet our technology has enabled us to view from the outside. Yet these images deceive. The blue marble offers the light side of the globe, leaving much to shadow. When we reify the Earth as a radiant sphere we seldom include the thing that makes it unique, its drenched and fragile atmosphere. The Earth is more extensive than its rocky surface. We ignore what we cannot see: the majority of the planet's matter, locked below our feet at the unreachable core. We know more about the surface of Mars than we do about Earth's interior.

Images from space make the globe seem solitary. Yet it has intimacy with the Moon, the satellite that churns its waters and makes its crust bulge, the piece of itself that circles Earth while altering Earth's own circling. Our planet spins on its axis while spinning around a spinning star, the source of warmth and light. The Sun meanwhile whirls within an arm of a rotating galaxy that hurtles through space. The solar system in motion is a helix, not a placid series of concentric circles (the Ptolemaic cosmos with the Sun at its center rather than the Earth). The Sun conveys its planets through space like buzzing children, their paths galactic loops. Spirals within spirals, nothing is still.

Maybe the Anthropocene is a circling, and we are never going to know when it began.

- 

Neither texts nor stones are timeless, despite what poets say. Both are so full of time that instead of lines and stadial history we get confounding spirals, aggregations of difference that intensify some stories and some climates and spin away from others. Floods of carbon or water may be arriving soon, just as they crashed in the primordial days of the Permian (which we did not cause) or Noah's flood (which we did). Maybe at the next recurrence, the one we are fashioning right now, we will pause and listen to Noah's wife as she sings her round and drinks her ale with those left to the weather. Maybe this time we will not leave an unloved world to drown.

- 

Disembarking from their houseboat after their long sojourn, Noah and his refugee family offer sacrifices to a preserving deity. After wiping Earth clean of all humans had possessed, God promises that he will never again send waters so planetary. The rainbow he hangs in the sky—a real bow, the kind that shoots you dead—is aimed toward heaven, not Earth. At his heart, not theirs. Is it a sign perhaps that he went too far? As God takes leave of those who have endured the flood he says to Noah, "And now farewell, my darling dear" ("Noyes Flood," line 375).

The last version of the Chester play dates from 1607, three years too early to be part of the 1610 Anthropocene. And yet I cannot help but hear in the play's closing words something like Prospero's recognition that to drown his books will bind them to the archive they deserve. When a dry box in which texts are cherished for their lined inscriptions is traded for a tempest of swirled water, an eddy where they might become something richer, stranger, then we possess an Anthropocene without origins, a flow without cuts, a history without bolted chests, a sea rich in opened arks, a cataclysm in which we do not leave those outside our walls to drown.

- 

*Refuge for some widened companionships, a difficult story of curves but not ends.*

## Notes

My sincere thanks to Tobias Menely, Steve Mentz, Jesse Oak Taylor, and especially Julian Yates for their impress on this essay, which is so much better for their care. I am also indebted to my frequent collaborator Lowell Duckert, without whom the vorticular frame of this chapter would not have been possible.

1. Noah Heringman, "Deep Time at the Dawn of the Anthropocene," *Representations* 129.1 (2015): 56–85, 58 (hereafter cited parenthetically in text).

2. For an eloquent examination of the conceptual challenges of this convergence of human and nonhuman history, and its opening up of new ways of reading the deep past through intensified attention to species effects, see Dipesh Chakrabarty, "The Climate of History: Four Theses," *Critical Inquiry* 35.2 (2009): 197–222.

3. For more on geology as a science that arises in the nineteenth century to read nature's "book of stone," see Bronislaw Szerszynski, "The End of the End of Nature: The Anthropocene and the Fate of the Human," *Oxford Literary Review* 34.2 (2012): 165–84. Szerszynski writes of geology's "linear but contingent deep history, the constant movement from surface differences to deep unities" and the "page-and-writing structure of strata and fossil through which Earth seems to write its own history" (178–79).

4. Such vorticular modes of reading owe much to the contributors to Jeffrey Jerome Cohen and Lowell Duckert, eds., *Elemental Ecocriticism: Thinking with Earth, Air, Water, and Fire* (Minneapolis: University of Minnesota Press, 2015).

5. See Dana Luciano, "Speaking Substances: Rock," *Los Angeles Review of Books*, April 12, 2016, https://lareviewofbooks.org/essay/speaking-substances-rock. I am also thinking here of the time-bending Anthropocene work of Srinivas Aravamudan, "The Catachronism of Climate Change," *Diacritics* 41.3 (2013): 6–30.

6. Mick Smith, *Against Ecological Sovereignty: Ethics, Biopolitics, and Saving the Natural World* (Minneapolis: University of Minnesota Press, 2011), 62 (hereafter cited parenthetically in text).

7. My thinking about topography, linearity, exteriority, and whorls owes much to the work of Tim Ingold; see especially *Being Alive: Essays on Movement, Knowledge and Description* (London: Routledge, 2011).

8. Timothy Morton, "How I Learned to Stop Worrying and Love the Term Anthropocene," *Cambridge Journal of Postcolonial Literary Inquiry* 1.2 (2014): 257–64, 264 (hereafter cited parenthetically in text).

9. The quotations in this paragraph are from "Noyes Flood," in *The Chester Mystery Cycle*, ed. R. M. Lumiansky and David Mills (Oxford: Oxford University Press, 1974), lines 225–28.

10. For the palimpsest, a figure for temporality, as thick, explosive, and simultaneous (an idea derived in large part from the work of Bruno Latour and Michel Serres), see Jonathan Gil Harris, *Untimely Matter in the Time of Shakespeare* (Philadelphia: University of Pennsylvania Press, 2009).

11. Michel Serres with Bruno Latour, *Conversations on Science, Culture, and Time*, trans. Roxanne Lapidus (Ann Arbor: University of Michigan Press, 1995), 58–59. Holding to a "chaotic theory of time," Serres argues that although time is often posited to flow linearly, like a laminar river, such a straightening figure ignores "counter-currents" and "turbulences" (58), so that "every historical is likewise multitemporal, simultaneously drawing from the obsolete, the contemporary and the futuristic. An object, a circumstance, is thus polychromic, multitemporal, and reveals a time that is gathered together, with multiple pleats" (60).

12. Ian Baucom, "History 4°: Postcolonial Method and Anthropocene Time," *Cambridge Journal of Postcolonial Literary Inquiry* 1.1 (2014): 123–42. Baucom usefully thinks through the necessity of periodization as part of his project.

13. *Oxford English Dictionary Online*, s.v. "sediment."

14. The term "uniformitarianism" was coined by William Whewell in 1832 as he reviewed Charles Lyell's *Principles of Geology* (1830–33), the book through which Hutton's ideas became widely known.

15. James Hutton, *Theory of the Earth, with Proofs and Illustrations*, ed. Sir Archibald Geikie (1785; repr., Delmar, N.Y.: Scholars' Facsimiles and Reprints, 1997), 3:46. On geology as a contemporary reading practice, see Noah Heringman, *Romantic Rocks, Aesthetic Geology* (Ithaca: Cornell University Press, 2004).

16. These words are the most famous and widely quoted that Hutton composed; they appear in *Abstract of a Dissertation Read in the Royal Society Edinburgh Concerning the System of the Earth, Its Duration, and Stability* (Edinburgh, 1785), 28.

17. John Playfair, *Illustrations of the Huttonian Theory of the Earth* (Edinburgh, 1802), 212, 117.

18. For the intimate relation between Noah's wife and the animals she brings aboard the ark, see Lisa Kiser, "The Animals in Chester's *Noah's Flood*," *Early Theatre* 14.1 (2001): 15–44.

19. The chart may be accessed via the ICS website: http://www.stratigraphy.org/index.php/ics-chart-timescale.

20. The official recommendation of the Anthropocene Working Group is for a start date of 1950, coinciding with the aftermath of the detonation of nuclear bombs and the embedding of radioactive materials in the geological record. On the stakes of determining when to instigate the Anthropocene—and what to call it—see especially Stacy Alaimo, "Your Shell on Acid: Material Immersion, Anthropocene Dissolves," in *Anthropocene Feminism*, ed. Richard Grusin (Minneapolis: University of Minnesota Press, forthcoming); Donna Haraway, "Anthropocene, Capitalocene, Plantationocene, Chthulucene: Making Kin," *Environmental Humanities* 6 (2015): 159–65; Steve Mentz, *Shipwreck Modernity: Ecologies of Globalization, 1550–1719* (Minneapolis: University of Minnesota Press, 2015), ix–xxiii; Jan Zalasiewicz et al., "Colonization of the Americas, 'Little Ice Age' Climate, and Bomb-Produced Carbon: Their Role in Defining the Anthropocene," *Anthropocene Review* 2.2 (2015): 117–27.

21. The first three quotations in this section are from "Noyes Flood," lines 225–28.

22. Ibid., lines 239, 242–45.

23. Eric Gidal, *Ossianic Unconformities: Bardic Poetry in the Industrial Age* (Charlottesville: University of Virginia Press, 2015), 6 (hereafter cited parenthetically in text). Gidal writes of creating a "hybrid taxonomy" of these unconformities to "register and reflect upon the social and spatial disruptions of industrialized modernity and the stratigraphic consciousness of geological deep time" (12).

24. "Vessel of modernity" is from Charles Withers, *Placing the Enlightenment: Thinking Geographically About the Age of Reason* (Chicago: University of Chicago Press, 2007), 209, cited approvingly by Gidal, who applies it especially to the steamship (*Ossianic Unconformities*, 10).

25. Gidal admits a confluence in the tropes but writes that the steam-powered ship is the conveyor of a reshaped "geopoetic imaginary" (*Ossianic Unconformities*, 161). Stone and water converge, I think, in desiring narratives of intensification over those of rupture. I am not willing to unalign the Roman trireme, the longboat, the schooner, and the freighter.

26. I am thinking here of Steve Mentz's notion of a wet or briny text in *At the Bottom of Shakespeare's Ocean* (London: Continuum, 2009).

27. See Mentz's idea of the Naufragocene, or the Anthropocene as the Age of Shipwreck, in *Shipwreck Modernity*.

28. John Mitchell made this point forcefully at the "Ecological Resilience" seminar of the Shakespeare Association of America's annual conference in New Orleans in 2016.

29. The work of the historian R. R. Davies is essential here. See especially *The First English Empire: Power and Identities in the British Isles, 1093–1343* (Oxford: Oxford University Press, 2000).

30. Interestingly, *Beowulf* contains the triad of mariner, stone, and text that marks Ossianic literature—and does so without Homer, thereby pushing hard at what Gidal's textual unconformity might signify. The recurring figures of the navigator, the stone that commemorates the warrior, and the text that memorializes them both is central to Gidal's reading of the Ossianic recording of Homer within a "brave new world of industrial modernity" (*Ossianic Unconformities*, 159), but there is also an untimely episode that unfolds in *Beowulf*, when the dying king imagines a future navigator spotting his tomb and relating his life.

31. Julie Orlemanski, "Scales of Reading," *Exemplaria* 26 (2014): 215–33, 217 (hereafter cited parenthetically in text).

32. In this volume, Mentz writes of a four-hundred-year-old Anthropocene: "The 1610 Anthropocene takes the latest claim for radical newness and submerges it back into history, with all of history's messiness and swirl." For a counterargument to the plausibility of 1610 as a marker, see especially Zalasiewicz et al., "Colonization of the Americas."

33. Jan Zalasiewicz et al., "Are We Now Living in the Anthropocene?," *GSA Today* 18.2 (2008): 4–8, 4.

# 2.
# Enter Anthropocene, Circa 1610

*Steve Mentz*

**Scene 1. Enter Anthropos solus, reading a book**

*He looks like one of us. Just the slightest hitch in his step, as if he's a little bit older than he appears. Academic tweed drapes his shoulders. The stage light glares off the crown of his bald head. He walks toward center stage slowly, eyes on his book, the pages of which he turns slowly with a look of concentration. It's hard to see the title, but we know it's a copy of* The Tempest.[1] *What else would it be?*

*He stops and looks up, startled. He says, "It's all my fault." Then he looks back down at the text and begins furiously turning pages. He finds a passage, puts a finger on the exact spot. He beams as if he has found what he's been looking for. "It's all my fault!" he insists. He brings the book closer to his eyes, stares at it a while, and then looks up at the audience. He speaks deliberately, with a hint of confused resignation that shades into perverse triumph, as if no one else is willing to pose this astounding question.*

*"It's ... all ... my ... fault?"*

*As he slowly walks off the stage, an article from the journal* Nature *falls out of his book. The title page shows Simon Lewis and Mark Maslin's essay "Defining the Anthropocene" (March 2015). Anthropos has highlighted the date 1610.*

The core of my argument about the 1610 Anthropocene as an incitement to revise our reading and interpretive practices can be compressed into the three different marks of punctuation that conclude the four-word sentence repeated by Anthropos: . ! ? First he declares his guilt with a period. Then he shouts an exclamation. Finally he questions his own ideas. As with Anthropos, so the

Anthropocene, both in its 1610 version and more broadly: we state it, shout it, and are left questioning. The epoch is not one thing but many, even when attached to a precise date. Each proposed punctuation forecloses necessary alternatives. We need all the options together. Pluralize the Anthropocene!

The controversial postulation by geographers Lewis and Maslin of 1610 as a possible golden spike opens up interpretive paths for humanities scholarship in this new and catastrophic Age of Man. The Anthropocene poses a challenge for narration and erects barriers to understanding texts that we once thought we knew. The period, exclamation point, and question mark each punctuate a distinct response to the challenges of our new epoch. Some thinkers rely on the declarative periods, sticking to bare facts: carbon in the atmosphere, temperature, sea level. Others address the Anthropocene with horrific exclamations. Still others linger amid messy questions. The three punctuation marks together tell a composite story: Anthropos is here. He's made everything bad! Now what do we do?

Current debates about the nature and historical timing of the Anthropocene build on geophysical techniques that have enabled the reading of stratigraphic "writing" in nonhuman media such as ice cores, tree rings, and fossil records. As the introduction to this volume shows, this new kind of reading resonates with and is subject to the same critiques as the reading that occupies humanities scholarship. In trespassing on scientific practices as a scholar of early modern literary culture, I proceed with a desire to engage the narrative forms of Anthropocene stories in many media. The three marks of punctuation that follow Old Man Anthropos's repeated sentence comprise narrative markers and senses of possible endings. The affect-less period claims the neutral voice of scientific observation. The exclamation point signals tragic ecstasy, which recalls that, since Sophocles, tragedy has been the narrative of highest prestige in Western literary culture. The question mark gestures toward futures structured through hybridity. The multiplicity of these boundary markers suggests that they need to be read together as narrative makers as well as separating marks. The triadic emphases of period, exclamation point, and question mark create a messy but partially legible set of Anthropocene narratives.

Opening up these three ways of marking Anthropocene stories recalls the genre varieties that, since Aristotle, have been signature analytical categories for the literary humanities. Marks of punctuation, like stratigraphic signatures, signal borders between disparate things, but genre definitions both label distinct entities and are themselves tools for creating new hybrids. Reading in the

Anthropocene defines itself through these two poles: marks that signal endings or boundary points, and the hybrid fecundity of mixtures that promise new kinds of stories. The crucial thing about these models is that they assume forms that literary scholars find legible. The rhetorical techniques of literary scholarship can make sense of these forms by relating them to our shared experience of narrative history. As an early modernist, my expertise emerges from a polygenre literary culture whose watchwords have been *imitatio* (we must make our stories like other stories!) and *contaminatio* (we must combine unlike narratives in order to make new ones!).[2] The language of genre imitation and contamination provides a powerful tool for Anthropocene reading. Caroline Levine's reconsideration of the strategic value of genre-based critical analysis suggests the continuing currency of such hermeneutic techniques.[3] Genre hybridity can overwrite punctuation's rage for closure.

The 1610 date suggests that the Age of Man may be less new than some exclamation-point narratives insist. History, like coal, agriculture, industrial carbon, and nuclear fission, produces a legible record of human marks on nonhuman nature. My admittedly messy suggestion in this chapter is that we read all the signs at the same time. We need not one Anthropocene but many, each sliding into and disrupting the others. It will not be easy to adapt our reading habits, since the basic practices of academic reading include separating and delineating differences or, we might say, marking punctuation.

The conceptual imperative of an Anthropocene requires that we reopen interpretations of both new and familiar texts: we need not just the bare periods of human causation, but also exclamatory zeal and questioning plurality. I make the early modernist's case for the 1610 Anthropocene, but I do not want that date to exclude 1950 or 1964 or 1712 or 2000 or 10,000 B.C.E. I propose that we read all Anthropocenes at once, while remembering that each one is revisable and provisional. There are many marks in the many human and nonhuman records we must consult—but no fixed marks!

Reading amid all this disorder poses challenges. Asking *The Tempest* to help in this process may seem logical—who is more central to Western reading practices than Big Will?—but I want specifically to emphasize the play's hybrid tragicomic genre and its varied interpretive and creative histories. *The Tempest* is not a singularizing mark or boundary but a pluralizing genre maker. The play's dual status as marker and maker gives it special meaning at the onset of many Anthropocenes, though it is hardly the only literary text to have such resonance. Opening our imaginations to changes that we cannot fully accommodate turns out to be exactly what both the Anthropocene and this hypercanonical play ask of us.

**Scene 2. Enter Anthropos, carrying a rugby ball**

*He tosses the ball from hand to hand. He's enjoying himself now, killing time before other players come to join the game. He reads aloud quotations from his book, each time looking up as if someone is about to arrive and receive a pass. In the last act of The Tempest, the two lovers are playing chess. "Sweet lord, you play me false," Miranda says. "No, my dearest love, / I would not for the world," Ferdinand replies. "Yes, for a score of kingdoms you should wrangle, / And I would call it fair play," she exclaims in rebuttal (5.1.172–75). Near the play's end, the happy couple plays—but to our surprise, they cheat and "wrangle." They pretend not to be worried about the story that they live in, the wizard-father who controls them, and the brave new future that awaits them. But we know—we can hear in their vocabulary—that they teem with anxiety.*

The Anthropocene poses a narrative problem. Narrative problems are not the only ones we face today—learning to retell and reinterpret stories of nonhuman nature will not materially move us into the postcarbon future we need—but narrative problems are what humanists are trained to solve, or at least to trouble. We keep passing the ball, it stays meaningful, and teams form around it.

Nothing entices and terrifies Old Man Anthropos more than chaos. An age of ecocatastrophe is an age in which all marks and all punctuation appear broken. The central question for the humanities today, to my mind, is how we can best salvage form from inside disorder. I recognize, as the contribution to this volume by Jeffrey Cohen states eloquently, that golden spikes, histories of rupture, and genre codifications all create falsehoods.[4] All marks and divisions distort and disrupt. But I remain drawn to brief appearances of order. I do not expect or want to get all the way to conceptual tidiness, but I value the chase and its transient marks. The principle of post-sustainability insists that change, not stasis, is the default structure, but amid disorder we still seek form.

The frustration of not being able to read provisional structures as anything other than provisional speaks to the dilemma of interpreting the Anthropocene and its many proposed start dates. It might be possible to let go of period markers entirely, as Cohen suggests. But I also wonder, perverse reader that I am, about what might happen if we indulge in precisely the opposite stance.

What if, instead of rejecting all periodizations because they cannot unlock the interpretive treasures that they promise, we accept all of them, provisionally? What if all the Anthropocene's origin stories—the birth of agriculture twelve thousand years ago, ecological globalization in early modernity, coal, the steam engine, petroleum, nuclear power—are meaningful, but none conclusive?

The version-and-inversion game that attempts to choose between different articulations of environmental and cultural history might seem a technical matter of interest mainly to super-subtle premodern historicist scholars. The key point may be simply the need for plural conceptions of historical change and difference. The year 1610 remains a flash point; it may be that over the long course of history, too much has already been made of this particular date—including, perhaps, yet another chapter that fingers Shakespeare and *The Tempest* as an inflection point. I want to escape doctrinaire versions of modernity, the human, and the Renaissance, though I am gambling that these things can be reshaped from inside. I want 1610 to mark an Anthropocene defined through resistant rereading rather than recapitulations of the story we all think we know. We must make the old new—which, perhaps, means that even on a familiar stage, Anthropos remains stuck in the middle of things. He remains subject to changes he (and we) cannot see in full.

**Scene 3. Enter Anthropos, carrying an umbrella**

*He looks scared. He reads in a quavering voice from the play's opening scene. He does not start at the beginning and read through to the end. He reads only the lines in which the Master and the Boatswain use technical sea terms. After each line, he looks up nervously at the audience, checking for reactions. He's looking for a way to survive the storm: "Fall to't yarely or we run ourselves aground. Bestir, bestir!" (1.1.3–4). "Yare! Yare! Take in the topsail. Tend to the master's whistle!" (1.1.6–7). "Use your authority! If you cannot, give thanks you have lived so long, and make yourself ready in your cabin for the mischance of the hour, if it so hap. Cheerly, good hearts.—Out of our way, I say!" (1.1.23–26). "Down with the topmast! Yare! Lower, lower! Bring her to try with main course" (1.1.33–34). "Lay her a-hold, a-hold! Set her two courses off to sea again! Lay her off!" (1.1.48–49). "We split, we split!—Farewell my wife and children!—Farewell brother! We split, we split, we split!" (1.1.60–62).*

For Anthropos, language is a life buoy. The magic words he wants will keep the ship off the rocks, or at least they will cue and organize the expert labor that might be able to do so. His words figure the community of working sailors as a unified whole, brothers who recall absent families, "good hearts." The ship is the state. It is also the body, both politic and organic. Shipwreck tells a radical story about the shattering and possible remaking of forms of order. What new things might be possible after "we split"?

Thinking about catastrophic openings and *The Tempest* returns my focus to the proposed date of 1610, which concentrates my mind by landing square in my own period of literary expertise. The story of a 1610 Anthropocene intimates that the storms of early modern globalization cracked and rebuilt the world's ecology and economy. This story is worth retelling, though it comes with risks. Here it is now:

Old Man Anthropos has a new date. I do not believe in magic numbers, but this one has got me thinking.

The Anthropocene, or the Age of Man, has become the word of the day. Making a bid to replace the Holocene, or the Age of the Present, as the scientific term for the geological era in which we live, the Anthropocene has caught the attention of scientists, scholars, artists, poets, theorists, and the general public. As humanist and posthumanist critics explore the era's implications, scientific debate continues about its precise nature. The question of origins remains vexed: when did the Age of Man start? Lewis and Maslin argue that the clearest geological markers of human influence on the global climate appear in 1964 and 1610.[5] The twentieth-century date reflects the peak of radioactive particles in the atmosphere, which subsequently declined after the Limited Nuclear Test Ban Treaty of 1963. But the earlier date catches this Shakespeare professor's eye: 1610 is three years after the founding of the Jamestown colony and one year before the first staging of *The Tempest*. Amid the glories of the English Renaissance sits an ecological spike. When Sir Walter Raleigh graced Queen Elizabeth's court and Shakespeare's dramas were first staged, our Anthropocene nightmare began. Or so goes the story. Does it end with a period, an exclamation point, or a question mark?

One problem with this date is that believing in it too much might support the swerve-into-modernity narrative, which has received some scholarly pushback.[6] But Lewis and Maslin do not base their claim for a 1610 spike on newly recovered manuscripts of Lucretius or the Baconian trio of print, gunpowder, and the compass. Instead Lewis and Maslin state that 1610 marks "an unambiguously permanent change to the Earth system" generated by the ecological mixing of the Americas with Europe, Africa, and Asia (177). The starkest consequence of this mixing from a human perspective was death on an unprecedented scale, primarily among Native Americans. Estimates vary, but the New World may have experienced the loss of nearly fifty million souls during the century of first contact, out of an estimated precontact population of roughly sixty to sixty-five million.[7] No period in recorded history matches this death toll in scale. The massive die-off of the human population and subsequent "cessation of farming and reduction in fire use" led to the "regeneration of over 50 million hectares of forest, woody savanna and grassland" (175). The open vistas of the New World were not destiny's gift to European settlers. The empty landscapes were visible evidence of the Anthropocene. The Age of Man is an Age of Death.

Lewis and Maslin name the 1610 date the "Orbis" spike (from the Latin for "world") because its drivers are global: the worldwide movements of human and nonhuman populations, as well as other factors, including "colonialism [and] global trade." As Dana Luciano notes in *Avidly*, this spike describes an Anthropocene that emerges not from industrial expansion but through such phenomena as the "concurrent history of the Atlantic slave trade."[8] The 1610 Anthropocene represents the early stages of what we now call "globalization." What might a global Anthropocene that shares its era with Shakespeare and Pocahontas mean?

If we hazard that the Age of Man started in 1610, we can reinterpret familiar words. Listening with Anthropocene ears, we hear old things in new ways. The magician's voice has changed. On the upper stage stands Prospero enrobed, singing out magnificent poetry in the voice of Gandalf and Magneto:

> Ye elves of hills, brooks, standing lakes and groves,
> And ye that on the sands with printless foot
> Do chase the ebbing Neptune, and do fly him
> When he comes back. (5.1.33–36)

If we have ears to hear, we realize Shakespeare's wizard sings destruction and the depopulation of the world. He creates and revels in ecological disorder:

> I have bedimmed
> The noontide sun, called forth the mutinous winds,
> And 'twixt the green sea and the azured vault
> Set roaring war....
> [G]raves at my command
> Have waked their sleepers, ope'd and let 'em forth
> By my so potent art. (5.1.41–44, 48–50)

Enter Anthropocene

Whether in Milan or on a magic island, Duke Anthropocene presides: enchanting, indulging, releasing, destroying. His voice is not the only one to which we should listen—I find more hope in shipwrecked sailors, lovelorn poets, and disoriented pilots—but since we have been listening to him for so long, it might be time to reconsider what he is saying. Listening with Anthropocene ears risks the academic sin of anachronism, but that price is worth paying to reorient how we understand the abiding force of early modern literature in the present.

Resinging a Renaissance Anthropocene in blank verse suggests some new things about this increasingly popular term.

The 1610 Anthropocene means that death, not heat, is humanity's primary historical driver. We are not just making the world warmer but making it deadlier. Carnage tells an exclamation-point narrative, protesting against cruel modernity. Thomas Pynchon nailed this one back in the early 1970s, writing *Gravity's Rainbow* in his beach pad in southern California:

> This is the world just before men. Too violently pitched alive in constant flow to ever to be seen by men directly. They are meant only to look at it dead, in still strata, transputrefied to oil or coal. Alive, it was a threat: it was Titans, was an overpeaking of life so clangorous and mad, such a green corona around earth's body that some spoiler had to be brought in before it blew the Creation apart. So we, the crippled keepers, were sent out to multiply, to have dominion. God's spoilers. Us. Counter-revolutionaries. It is our mission to promote death.[9]

Pynchon's vision of humans as death promoters speaks to many Anthropocenes at once. He glosses the efforts of early humans in exterminating large mammals and Neanderthal rivals in his reference to the "spoiler" who achieves biblical "dominion." The 1945 of *Gravity's Rainbow* also puts its finger on the rocket trigger of the nuclear Anthropocene of Alamogordo and Hiroshima. Pynchon's description of coal and oil as "transputrefied" Titans encompasses a third possible Anthropocene, driven by the Industrial Revolution. For this expansive postmodern writer, all Anthropocenes appear possible, and all are marked by destruction.

The 1610 Anthropocene de-anthropocentrizes Pynchon's narrative by suggesting that the most consequential historical and ecological forces in the Age of Man have been inhuman viruses, not human industry. Smallpox and influenza cleared the New World for colonization; malaria made its tropical regions ripe for transatlantic slavery. This inhuman globalization connects with Jason Moore's notion of a 450-year-old Capitalocene as "a way of organizing nature."[10] Moore's project brings the nonhuman into capitalism through "a world history in which nature matters not merely as consequence, but as constitutive and active in the accumulation of abstract social labor" (84). Ecomodernity is not only a human story.

The 1610 Anthropocene means that the key motivation of the environmental changes our species created was a desire for global connection and wealth, not just industrial production or population growth. There are many ways to blame, aggrandize, or describe the globalizing energies of early modern expansion. In addition to the canonical Anthropocene and newer Capitalocene, I seek space for the Homogenocene, an Age of Increasing Ecological Sameness; a Thalassocene, or Age of Global Oceans; and—my real favorite—the Naufragocene, the Age of Shipwreck.[11] Each 'cene jostles the others; each connects and disconnects. Telling them all together risks chaos but brings to the surface stories that reveal how entangled the "human" and the "natural" have always been.

The 1610 Anthropocene used to be called the Columbian Exchange, but that term is too reminiscent of "great man" theories of history.[12] Old Man Anthropos may have started it, but he has never been in control. A better phrase, "ecological globalization," properly takes the soup out of human hands. We are in it, not cooking it.

Reconsidering the 1610 Anthropocene through both capitalist expansions and more-than-human confrontations helps emphasize that the core story—the story that still needs retelling and that meaningfully precedes the supposed modernity of the past half millennium—concerns the production of hybrids through the collision of unlike worlds. Hybrid stories emerge through genre imitation and contamination; they require multiply punctuated tellings. Creating new combinations is not just a 1610 question, even if some forms of hybridity blossomed during that period. Hybrid production typifies human cultural history, from Neolithic art to postmodern architecture. Bruno Latour has given us a robust language for hybridity, but our best guide may be Caribbean poet and theorist Édouard Glissant, whose idea of "relation" promises "a new and original dimension allowing each person to be there and elsewhere, rooted and open, lost in the mountains and free beneath the sea, in harmony and in errantry."[13] That is the way to navigate storms, in or beyond the Anthropocene. Harmony and errantry, together.

The 1610 Anthropocene takes the latest claim for radical newness and submerges it back into history, with all of history's messiness and swirl. A four-hundred-year-old Anthropocene promises an unstable future, one in which we need to recall our disorienting past. If human civilizations have always been environment makers, the mutual implication of human and nonhuman forces may not be so new after all. It turns out that this latest thing is also an old thing.

To recast a familiar phrase that has new resonance in an age of rising global temperatures: the past is not dead. It's just getting warmed up.

**Scene 4. Enter Anthropos, staring at his iPhone**

*The phone rings, but he's too startled to answer it. He stands still, listening to his ring tone: Neil Young's "Like a Hurricane." A storm is coming. Anthropos cups a hand to his ear. What's that he hears? Singing? Or curses? A voice snarls offstage:*

> As wicked dew as e'er my mother brushed
> With raven's feather from unwholesome fen
> Drop on you both. A southwest blow on ye
> And blister you all o'er. (1.2.322–25)

*Anthropos cringes. He's afraid of this sort of language and afraid of the witch-mother that Caliban's words promise. He wants to drain the fens and shelter himself from blistering winds. He silences his phone, and the cursing stops. He listens in the quiet for a countersong. It too comes from offstage, but from the opposite side. The voice of a young woman soothes his fears:*

> Work not so hard. I would the lightning had
> Burnt up those logs that you are enjoined to pile!
> Pray set it down and rest you. When this burns,
> 'Twill weep for having wearied you. My father
> Is hard at study; pray now, rest yourself.
> He's safe for these three hours. (3.1.15–21)

*To whom should Anthropos listen? The cursing monster who knows his way around the island, or the beautiful daughter who only wants him to be comfortable while the forest burns? To whom should we listen?*

Storms shape scenes, including Anthropocenes. We know that Shakespeare's most famous play circa 1610 is named *The Tempest*, but since its opening scene draws on descriptions of New World storms, perhaps it should have been called *The Hurricane*. Might that be the magic word after all? Might there be new stories spinning off from that vortex?

Hurricanes were rarely encountered by Europeans before the early modern period. First described by Columbus's second fleet in 1495, these distinctive New World storms struck European imaginations with the destructive novelty of the 1610 Anthropocene.[14] Like today's ecotheorists, early modern sailors and writers struggled to interpret alien meteorological phenomena. Sailors and scientists wrote explanatory narratives. Preachers railed in lamentation. Playwrights and epic poets crafted intricate and mysterious questions. These responses to an early modern Anthropocene gesture toward a narrative and hermeneutic multiplicity that challenges our interpretive skills. Against Old Man Anthropos's rage for endings and punctuation, vorticular storms and stories offer plurality and constant generation.

The nonhuman figures of *The Tempest*, Ariel and Caliban, have served European literary culture as mobile symbols of difference and change since Shakespeare's play was first performed shortly after 1610. Frank Kermode's Arden 2 edition (1954) advances the now-prominent argument that connects the play to the New World expansion that Lewis and Maslin name the Orbis spike.[15] Americanist Leo Marx's *The Machine in the Garden* (1964) extends Kermode by treating Shakespeare's play as an origin point for English-language writing about the Americas.[16] Imagining Prospero as a representative of European expansion and technological change has become a familiar gesture among postcolonial Shakespeareans.[17] The global legacies of the spirits of wind and earth, however, have been even more generative.

The diminution of Prospero and the reconsideration of the subaltern figures of Ariel, Caliban, and Miranda have orchestrated the vast surge of revisions of *The Tempest*. Chantel Zabus argues in *Tempests After Shakespeare* that this play represents the "interpellative dream-text" of seventeenth-century English literary culture.[18] That dream text was subsequently transformed with the cultural upheavals of each successive age. Postcolonial scholars, including Peter Hulme, have explored the vast reach of the play and its symbolic refigurations.[19] Martiniquean playwright Aimé Césaire concludes his version with Caliban's voice: "FREEDOM HI-DAY! FREEDOM HI-DAY!"[20] The Uruguayan intellectual José Enrique Rodó's essay *Ariel*, first published in 1900, imagines Latin American culture caught between the positivist Caliban and the spiritual Ariel. Cuban intellectual Roberto Fernández Retamar extended the *Tempest* metaphor in an essay on Caliban in 1973.[21] For these and other artists, from poet W. H. Auden during World War II to filmmaker Peter Greenaway in 1991, the play has generated new texts in new eras.[22]

The Anthropocene itself may also be generative even in its cataclysms. Returning yet again to *The Tempest* as both urtext and pretext for ideas about global culture during and after Western colonialism risks reifying old-fashioned ideas about universal genius. Anthropocene reading requires something new. We must revive texts we once thought we understood by listening to them with ears pricked by guilt, nostalgia, and looming ecocatastrophe. Inverting the play's hierarchies so that Caliban, Ariel, and Miranda topple Prospero may not do enough to dethrone Old Man Anthropos. Even imagining his words and his power bringing forth a disastrous Age of Death may not finally release his brutal grip on our cultural history. Anthropocene reading asks for renewed attention to the marks of punctuation that separate epochs and narratives from each other and also to genre hybridities that create new things with old tools. The period-punctuated stories of *The Tempest*, from Caliban's prefiguration of the Atlantic slave trade to Prospero's anticipation of Enlightenment science, seem clear. What happens if we add to them the question marks of Ariel's elemental force and Miranda's liberated desires? Hybrid possibilities flow out of Anthropocene relistening and retelling that encourage narratives to overflow their bounds. Turning Shakespeare's not-quite-named hurricane into a vorticular machine for generating new stories allows this old play to spin off new things.[23]

**Scene 5. Enter Anthropos solus, soaking wet**

*He shivers in the wind as he reads the epilogue in a whisper. The audience leans forward, straining to hear:*

> Now my charms are all o'erthrown,
> And what strength I have's mine own,
> Which is most faint. (Ep. 1–3)

*Is he speaking to us, or to the Earth? The Anthropocene, he seems to say, is a narrative of failure, not mastery.*

> *Gentle breath of yours my sails*
> *Must fill, or else my project fails,*
> *Which was to please. (Ep. 11–13)*

*He grows desperate, his voice breaking. He falls to his knees. Trying to please can be a frustrating project.*

> *Now I want*
> *Spirits to enforce, art to enchant;*
> *And my ending is despair,*
> *Unless I be relieved by prayer. (Ep. 13–16)*

*What does he mean when he asks for "indulgence" (Ep. 20)? Hasn't he had enough of that? Haven't we all had enough of indulging him?*

What if it does not work? What if the Anthropocene is not or cannot become a productive concept for the humanities? What if my *Tempest* fable leaves us caught inside canonicity, reading over and over the stories we already know? Perhaps the Anthropocene names an impossible thought, a task that humanities scholars cannot encompass, precisely because the two elements of this volume's subtitle, literary history and geologic times, are not finally compatible.[24] One powerful literary response to this dilemma might be tragedy, but as my tragicomic literary example suggests, I am seeking a way forward past tragedy, though perhaps with some tragic tears coming along for the ride.

At the risk of exposing my literary traditionalism, I respond to anxieties about narrative inadequacy by returning to genre and invention. I remain limited but also energized by my old-fashioned conception of the way novelties emerge over time, through recombinations of old materials. Genre innovations, from seventeenth-century tragicomedies like *The Tempest* to Thomas Hardy's Victorian novels, J. M. Coetzee's postmodern parables, and the wide range of other materials scholars discuss in and beyond this volume, emerge from remixing old things in new ways. We build with the tools we already have, but that does not mean we cannot build new things.

My final salvage operation for the admittedly arbitrary project of the 1610 Anthropocene returns to punctuation marks that establish endings and to genre hybrids that accumulate new forms. The Anthropocene represents

Enter Anthropocene

a shift in narrative emphasis, a redoubling of human self-awareness about our entanglements with nonhuman forms and forces. The shift is not new, though it has always been and still remains unevenly present. It has a 1610 inflection, among many others. The 1610 Anthropocene makes triumphs seem tragic and humanism a screen through which we can descry death. That critique of self-congratulatory cultural histories represents a meaningful reformulation of the period we still sometimes call the Renaissance. But such redirected readings are always only partial. Perhaps a full awareness of our physical and historical embeddedness in the world outside our minds and bodies would be unbearable, or perhaps it would be simple justice, or maybe just plain confusing.

I want all these marks not only because they represent indelible signatures in the Earth system but also because they can generate multiple stories. There is no erasure. The marks humans have made on our planet are fixed and lasting. But we can read these marks in multiple ways, even if we confine ourselves to hypercanonical scripts like Shakespeare's plays. The three ways to punctuate 1610 yield multiple hybrid narrative futures.

The period story flatly declares that culture is part of nature, and there is no humanity without countless nonhuman alien contributions, from the bacterial flora in our stomachs to the global play of winds and ocean currents. We need this story; we make use of it; we never quite know what it means.

The anguished exclamation narrative laments the Age of Death and the tragic awareness through which human consciousness storifies the world. Tales of justice and redemption pale before crimes that we can neither redress nor comprehend. This story hurts when we tell it.

For me, the disoriented question mark generates the best stories. Like voyages without destinations, these turnings provide ways to inhabit hybridity and change without rest. If it be not 1610, 'tis perhaps 1950. If it be not 1950, perhaps 1964. If it be not 10,000 B.C.E., yet it will come. The readiness is all.

Or so goes one story.

## Notes

1. Quotations from *The Tempest* are from Virginia Mason Vaughan and Alden T. Vaughan's revised edition (London: Bloomsbury, 2011), known as Arden 3 (hereafter cited parenthetically in text with act, scene, and line numbers).

2. On Italian genre theory in the sixteenth century, from which my terms are drawn, see Daniel Javitch, "Italian Epic Theory," in *The Cambridge Companion to Literary Criticism*, vol. 3, *The Renaissance*, ed. Glyn P. Norton (Cambridge: Cambridge University Press, 1989).

3. Caroline Levine, *Forms: Whole, Rhythm, Hierarchy, Network* (Princeton: Princeton University Press, 2015).

4. In addition to his chapter in this volume, see Jeffrey Jerome Cohen, *Stone: An Ecology of the Inhuman* (Minneapolis: University of Minnesota Press, 2015).

5. Simon L. Lewis and Mark A. Maslin, "Defining the Anthropocene," *Nature* 519 (March 2015): 171–80.

6. For responses to Stephen Greenblatt's *The Swerve: How the World Became Modern* (New York: Norton, 2012), see Tisan Pugh's collected Book Review Forum in *Exemplaria* 25.4 (October 2013): 313–70.

7. On the ecological consequences of early modern globalization, see Charles Mann, *1493: How the Ecological Collision of Europe and the Americas Gave Rise to the Modern World* (New York: Vintage, 2011).

8. Dana Luciano, "The Inhuman Anthropocene," *Avidly: Los Angeles Review of Books*, March 22, 2015, http://avidly.lareviewofbooks.org/2015/03/22/the-inhuman-anthropocene.

9. Thomas Pynchon, *Gravity's Rainbow* (New York: Penguin, 1973), 720.

10. Jason Moore, *Capitalism in the Web of Life: Ecology and the Accumulation of Capital* (London: Verso, 2015), 78 (hereafter cited parenthetically in text).

11. Steve Mentz, *Shipwreck Modernity: Ecologies of Globalization, 1550–1719* (Minneapolis: University of Minnesota Press, 2015), ix–xxiii.

12. Alfred Crosby, *The Columbian Exchange: Biological and Cultural Consequences of 1492*, 2nd ed. (New York: Praeger, 2003).

13. Édouard Glissant, *Poetics of Relation*, trans. Betsy Wing (Ann Arbor: University of Michigan Press, 1997), 34.

14. On hurricanes, see Peter Hulme, *Colonial Encounters: Europe and the Native Caribbean, 1492–1797* (London: Methuen, 1986). I expand on this point in Mentz, "Hurricanes, Tempests, and the Meteorological Globe," in *The Palgrave Handbook of Early Modern Literature, Science, and Culture*, ed. Evelyn Tribble and Howard Marchitello (London: Palgrave, 2017), 257–76.

15. See Kermode's introduction to William Shakespeare, *The Tempest* (London: Bloomsbury, 1954) (Arden 2), x–xciii.

16. Leo Marx, *The Machine in the Garden: Technology and the Pastoral Ideal* (Oxford: Oxford University Press, 1964).

17. See, among many others, Kim Hall, *Things of Darkness: Economies of Race and Gender in Early Modern England* (Ithaca: Cornell University Press, 1995); Hulme, *Colonial Encounters*; Stephen Greenblatt, "Learning to Curse: Aspects of Linguistic Colonialism in the Sixteenth Century," in his *Learning to Curse: Essays in Early Modern Culture* (New York: Routledge, 1991), 16–39.

18. Chantel Zabus, *Tempests After Shakespeare* (London: Palgrave Macmillan, 2002), 1.

19. See Peter Hulme and William Sherman, *The Tempest and Its Travels* (Philadelphia: University of Pennsylvania Press, 2000); Virginia Mason Vaughan and Alden Vaughan, *Shakespeare's Caliban: A Cultural History* (Cambridge: Cambridge University Press, 1993).

20. Aimé Césaire, *A Tempest*, trans. Richard Miller (New York: TCG Translations, 2002), 66.

21. José Enrique Rodó, *Ariel*, trans. Margaret Sayers Peden (1900; repr., Austin: University of Texas Press, 1988); Roberto Fernández Retamar, "Caliban: Notes Towards a Discussion of Culture in America," *Massachusetts Review* 15 (1973–74): 11–16.

22. W. H. Auden, *The Sea and the Mirror: A Commentary on Shakespeare's "The Tempest,"* ed. Arthur C. Kirsh (Princeton: Princeton University Press, 2005); Peter Greenaway, dir., *Prospero's Books* (1991).

23. On the vortex as environmental form, see Jeffrey Jerome Cohen, "The Sky

Above," in *Elemental Ecocriticism: Thinking with Earth, Air, Water, and Fire*, ed. Jeffrey Jerome Cohen and Lowell Duckert (Minneapolis: University of Minnesota Press, 2015), 107–16. Cohen also thinks with vortices in his contribution to this volume.

24. Benjamin Morgan's contribution to this volume makes this point through reading two early Hardy novels.

# 3.
# The Anthropocene Reads Buffon; or, Reading Like Geology

*Noah Heringman*

The writer of romance is to be considered as the writer of real history.
—William Godwin, "Of History and Romance" (1797)

What future reader, contemplating the tale of the Anthropocene, will exclaim, "Here is one of the most sublime romances, one of the most beautiful poems, that philosophy has ever dared to dream up!"[1] As noted by its current critics, many versions of this tale do not read like geology. Our future reader, however, might well note the struggle of "the species" against extinction as a topos recalling the fossil form of romance. The pathos of this quest, including its appeal for interspecies solidarity, might not be lost on her. She will be terrified, or amused, by the colossal scale of the action. Although there is some overlap, these topoi are not exactly what Jacques-Henri Meister had in mind when he greeted *Epochs of Nature* (1778) by Georges-Louis Leclerc, Comte de Buffon, with the exclamation quoted above. Rather, they have become newly salient for Anthropocene readers of Buffon, such as Claudine Cohen, Elizabeth Kolbert, and Jan Zalasiewicz. Scientific romance matters deeply for readers who wish to embrace or contest or define the Anthropocene. The issue is not whether the Anthropocene was "anticipated,"[2] although *Epochs of Nature* deserves a place in its intellectual history, a place most clearly indicated by the appearance of the first complete English translation of the work, currently being prepared by two leading theorists of the Anthropocene, Zalasiewicz and Jacques Grinevald, with other colleagues.[3] The issue is our current entanglement with scientific romance, a genre that is coming into its own in the Anthropocene.

Current interest in Buffon's *Epochs* marks certain continuities between eighteenth- and twenty-first-century innovations in the geological time scale. The forms of scientific romance that are proliferating around the Anthropocene differ substantially from those associated with the history of geology, however, both because there is a deeper divide now between research and popular science and because the survival of the species is perceived to be at stake. By reading Buffon with and against contemporary authors on geological time, especially Zalasiewicz and Kolbert (*The Sixth Extinction*), this chapter sets out to correlate the two episodes.

Buffon's first readers tended to use the term "romance" somewhat pejoratively, implying that his stylistic virtuosity was somehow at odds with philosophical rigor (as it would then have been called). H. G. Wells described his own works more positively as "scientific romances," but he did so under the aegis of a literature and science binary that mattered less for Buffon.[4] Anthropocene readers and writers might align themselves more willingly with popular science, a related idiom on which Buffon and Wells both left their mark. Twenty-first-century scientific approaches abound for addressing problems first posed imaginatively by Buffon and other self-styled readers of the "archives of nature," including John Whitehurst, Johann Reinhold Forster, and Georges Cuvier. The urgency of these problems today, however—such as the scale of geological time and the interdependence as well as the possible extinction of human and nonhuman species—dictates a revival of the literary form in which they were once posed. Anthropocene reading of early iterations of the geological time scale means attending to their fossil fuel subtexts—the "treasuries of nature" that Buffon locates in Earth's archive—and bringing the history of the form to bear on twenty-first-century narratives of geological time, especially the postcarbon romance of Anthropocene popular science. Even peer-reviewed science on the Anthropocene does not read like geology to some specialists because it diverges from strictly stratigraphic criteria or deviates into narrative, political and otherwise.[5] Shifting the verb from the middle voice to the active voice, we may also define "reading like geology" as the act of interpreting the fossil record, here especially looking at coal deposits as they would have appeared to eighteenth-century observers.

The far-distant future reader is one figure that Anthropocene popular science shares with the old geological romance.[6] Buffon not only gave his contemporaries their first glimpse of an outlandishly long Earth history, prompting cries of "Romance!"—he also insisted that human beings, though a recent arrival, would remain to witness a future equal in duration to the planet's past existence. Cuvier, who owed more to Buffon than he cared to admit, including the concept of extinction, added to the length of humanity's lease by declaring that the "revolutions" that had caused mass extinctions in the past were now greatly diminished in their force. Cuvier's systematic scrutiny of fossil fauna authorized his self-appointed title, "a new species of antiquary," but the antiquarian topos of ruin continued to designate the reading practice that would serve students of ancient life in these extended futures.[7]

Unlike their French precursors, those who would chronicle the Anthropocene are forced to contemplate the prospect of human extinction, but even so the figure of the future reader is indispensable to them. Recent scientific contributions weigh the relative chronostratigraphic merits of radioisotopes that will remain detectable fifty thousand or a hundred thousand years into the future. Clive Hamilton speaks casually of "what geologists in a million years will detect" ("Getting," 105), and Kolbert portrays Zalasiewicz as predicting confidently that the Anthropocene signature will remain legible to "even a moderately competent stratigrapher . . . at the distance of a hundred million years or so."[8] These figures have a strong heuristic foundation, but they are so extravagant that they must keep company with Wells's Eloi and Morlocks of 802,701 C.E. and even with the alien "species, after humans, 'reading' our planet and its archive" envisioned by Claire Colebrook.[9] This extravagance is not all. Pathos and wonder and even sympathetic identification must attend the magnificent catalog of past extinctions. The survival of the human species, dramatically set off against the history of extinctions and increasingly vast expanses of time, becomes a multispecies affair, fatefully entangled with fossil plants as well as living species. Finally, the geological reader depends on a Cyclopean cast of fossil characters for the essential distinction between geological and historical time, or any sense of scale. "[T]he human mind," Buffon observes, "loses itself in the expanse of duration much more readily than in that of space or number."[10]

Deep time has been claimed as the province of epic both "high" and "low," and Buffon's *Epochs* in particular has been dubbed an epic by Zalasiewicz, one of the "standard-bearers" of our newest epoch.[11] Popular science, with influences ranging from Lucretius to genre fiction, might well be classed as "middle epic" in this context. In this volume dedicated to Anthropocene reading, however, romance seems to me the strongest contender because it offers the improbable figures needed to situate human actors in deep time. The centuries-old critical project of distinguishing *The Odyssey*, a romantic epic, from the more heroic *Iliad* provides a substantial precedent for a discrimination of reading practices: *The Odyssey* is more romantic because it "mingles pathos with adventure"; it emphasizes "the qualities of vision (or 'imagination'), generous scale, and sustained description" that increasingly differentiated romance from epic in the Hellenistic period; and it displays the qualities of "marvel, risk, and triumphant adventure" that defined romance throughout the Middle Ages and into the early modern period.[12]

Neither Buffon nor Kolbert saw themselves as writing romance or even scientific romance period. I will argue, however, that both depend on romance tropes, including the distant future, extinction, multispecies agency, and extended scale. These four figures have shaped a historical kind of Anthropocene reading, the kind for which Buffon still matters, or, put another way, a reading that attends to fossil fuel subtexts. One common thread linking these figures is "marvel," or wonder, a quality that Richard Holmes has used to define the quest for natural knowledge in the age of Buffon and Cuvier.[13] This historical association helps to explain why *The Sixth Extinction* derives one of its main plot lines, the history of the concept of extinction, from the history of science. Wonder in a broader sense helps to account for the proliferation and popularity of Anthropocene narratives. Kolbert's historical engagement offers an alternative to the uncritical endorsement of "our" geological "superpowers" that contaminates some of these narratives,[14] because she ends with a dispassionate look at human extinction that emulates the alien gaze of Colebrook's "inhuman" post-Anthropocene reader. Scientific romance in the strict sense—an Anglophone tradition stretching from C. H. Hinton and Wells to Ronald Wright—often mounts its social critique on the strength of wonder, or on the dystopian sublime now commonly derided as "disaster porn," but it does so at the cost of its scientific truth claims. Wells insisted that his scientific romances traded not in the "actual possibilities of invention and discovery" but instead promoted a social vision through the "ingenious use of scientific patter."[15] Contra Wells and contra Aristotle on Empedocles, the scientific romance as initiated by Buffon and revived in the Anthropocene is fundamentally a narrative nonfiction form.

Reading popular science as romance may yield a critical purchase on a surprising variety of texts for the Anthropocene reader. What is often called "our romance with fossil fuels" is too seldom considered as a literary construction in the first place. Romance topoi blossomed around coal even before the Industrial Revolution, as my readings of Buffon and his contemporaries will show. Although it may contain "the first real expression of the twenty-first-century Anthropocene concept," as Zalasiewicz argues, *Epochs of Nature* is not manifestly "about" coal.[16] Coal does, however, mediate between geological and historical time in works by Buffon, Whitehurst, Cuvier, and other contemporaries, and it focalizes a larger narrative of resource extraction that impinges on them. This fossil fuel subtext exceeds "the traditional notion of 'context,'" as Fredric Jameson put it when he proposed "subtext" as a critical rubric.[17] But neither is it a set of social contradictions to be reconstructed under the hermeneutic lens of "symptomatic reading," as Stephen Best and Sharon Marcus somewhat skeptically term Jameson's critical practice.[18] This is because the fossil fuel subtext instantiates what Menely and Taylor (this volume) call a "historicity that exceeds the human social relation." Only the consumption of fossil fuels has rendered them historical. Reading for coal in Enlightenment geochronology is a strategy that can benefit from the denotative approach outlined by Elaine Freedgood and Cannon Schmitt (who use coal mining as their primary example),[19] but it also entails a subterranean passage through the non-carboniferous strata—geological and otherwise—that have made geological time salient and (nearly) intelligible. If we set the parameters of "our romance with fossil fuels" in this way, a contemporary text such as Simon Lewis and Mark Maslin's "Defining the Anthropocene" can yield a further inversion, a perspective on globalization as the subtext not of literature but of geology.[20] The revival of scientific romance in the Anthropocene generates rich possibilities for a literary history engaged in digging up what Falko Schmieder has called "the buried history of the Anthropocene."[21]

If the old romances of the Earth now read like geology after all, how will Anthropocene geology (our fossil remains, in this case, and not our stories) be received by our future reader? More important, can our understanding of the human and nonhuman subtext of the geological time scale (GTS) enable a more critical consumption of the popular science that literary academics increasingly depend on? This chapter outlines a program of "reading like geology" by delineating some of the romance topoi that flourished in Enlightenment geochronology and recovering the senses of "romance" mobilized by the original critics of these narratives, while also presenting evidence that a revival of this narrative form is under way. Jean-Étienne Guettard made it very clear that *Epochs of Nature* did not, in his view, read like geology (or geotheory, to adopt Martin Rudwick's more precise historical term). "More Buffonades, my dear Count, how much longer will you play Cyrano de Bergerac?" exclaims Guettard in a letter, alluding to both English and French literature in one exasperated breath. Citing Buffon's "hypothetical ideas" about the formation of planets and their long, slow extinction, Guettard charges him with getting "carried away into romance" and predicts caustically that his *romans* will "ravish the antechambers" and pass through ladies' toilets to be "devoured" by the "chambermaid" and to "amuse" the "lackey."[22] Johann Reinhold Forster calls *Epochs of Nature* a "beautiful romance/novel [*den wunderschönen Roman*]" in a less unfavorable, but still skeptical review.[23] Only a generation later, John Playfair goes one step further to deride all previous theories of the Earth as "a species of mental derangement."[24] Looking back on Buffon's reception today, Zalasiewicz identifies elements of the "visionary" and of "melodrama" in *Epochs of Nature*, but recognizes that these formal features do not diminish its stature as "the first science-based whole-Earth narrative."[25] By showing how *Epochs of Nature* does, in the end, read like geology, Zalasiewicz's reading of Buffon models a practice of foregrounding literary tropes in geological writing, not for the sake of undermining its truth claims (as we find in Guettard and Forster, as well as in some critics of the Anthropocene), but for the sake of promoting a historical understanding of geological time. Reading Buffon and his critics clarifies the nature of the Anthropocene intervention in the GTS.

The charge of romance, as both a narrative strategy and a wedge for critique, still circulates through the "science-based whole-Earth narratives" of the present, although we are less apt to dismiss these narratives for being accessible or popular, as Guettard did in his scathing letter (which was probably never sent). Contributions to Anthropocene science require a measure of accessibility to reach beyond specialist communities to the interdisciplinary readership of *Nature*, *Science*, or new journals such as the *Anthropocene Review*. Lewis and Maslin recognize explicitly that the "choice" of a start date for the new epoch implicitly "tells a story" about technology ("Defining," 177), a reflexive move that may account for the skeptical reaction to their work by some commentators. Certainly a perception that proponents of the Anthropocene are willing to instrumentalize the term in this way has led to critical response, perhaps especially in the U.S. geological community. Hamilton and Grinevald (among others) have challenged another Anthropocene story ("Was the Anthropocene," 62), William F. Ruddiman's "early Anthropocene" thesis, on the grounds that it merely retells the familiar tale of the Holocene.[26] Hamilton elsewhere critiques Lewis and Maslin too, but on the more categorical basis that the choice of a start date for the new epoch "can be like no previous one" ("Getting," 105). The members of the Anthropocene Working Group argue that, on the contrary, the boundary must be "dispassionately chosen, by the same manner in which all earlier stratigraphic boundaries were chosen."[27] In his own writing on Buffon, Zalasiewicz articulates a position close to Lewis and Maslin's on the importance of story, citing the interdisciplinary makeup of the working group (which includes a journalist and a historian, but no *romanciers*) and arguing the need for latter-day savants who can speak to the broader social implications of the Anthropocene.[28] Readability, however, remains a specialist criterion as well: the radionucleides that Hamilton takes for granted as "merely a signifier" ("Getting," 105) are not all that different from the stratigraphic "text" envisioned by Whitehurst in 1778 as "a language and characters equally intelligible to all nations."[29] Here is a more positive take on the accessibility that led Guettard to dismiss Buffon's *roman* as fodder for chambermaids and lackeys. As Jeffrey J. Cohen argues in this volume, "linearity" may be the shared limitation of the many Anthropocene stories, romantic and otherwise, that focus exclusively on locating the radical break with the past that it constitutes, both methodologically and substantively. Cohen looks to medieval romances, among other tales, to map instead the cyclical time of Anthropocene "melancholy."

Buffon's seventh epoch, the recent "time of man," practically guarantees a long-term future for the species through the mechanisms of human exceptionalism and resource superabundance, which were not controversial in the Enlightenment, and by situating a human actor in geological time, which was. Long-term future scenarios today are haunted, conversely, by the prospect of resource exhaustion and all it implies for the exploitation of nature. The narrative structure of situating a human actor in geological time, however, is not controversial today—quite the opposite. This narrative arc in Buffon is another precedent for the Anthropocene, in which the "history of the [human] species" takes on a special prominence.[30] To begin with superabundance:

> [Coal mines are] treasuries that Nature seems to have accumulated in advance for the future needs of large populations; the more human beings multiply, the more forests diminish: wood will no longer be enough to satisfy them, they will have recourse to these immense deposits of combustible matter, the use of which will become that much more essential with the increasing refrigeration of the globe; nonetheless, they will never exhaust them, because a single coal mine contains perhaps as much combustible matter as all the forests of one vast country combined. (*Époques*, 96)

Buffon echoes this theme of superabundance in the rhetorically powerful conclusion to his book, which holds out "treasuries of inexhaustible fecundity" against time's dark abyss (*Époques*, 220). These inexhaustible treasuries, together with what is only an incipiently secularized humanist ethos of stewardship, render it even possible that human life can be sustained as long as the Earth has existed—another seventy-six thousand years, if not more (213). Although seventy-six thousand years seems short by the standards of today's GTS, it was outrageously long to the faculty of the Sorbonne. More crucially, it is dizzyingly long compared to the few centuries it actually took to use up this "inexhaustible" wealth.[31]

Our precocious situation as future readers of Buffon's Earth is partly an effect of the history of science. One of the implications of the charge of "romance" leveled at Buffon by younger naturalists, such as Forster and Guettard, is that his natural philosophy was already obsolete. The same implication comes through in Cuvier's 1829 eloge on Jean-Baptiste Lamarck, whose theories he compares to "the enchanted palaces of our old romances."[32] The romantic qualities of obsolete science become increasingly clear with the passage of time, as when Playfair looks back to Thomas Burnet's and William Whiston's theories of the Earth, or when Mary Shelley's Professor Krempe dismisses the "ancient . . . fancies" of Paracelsus.[33] Yet obsolete science can prove to be stunningly correct, inspiring a more positive form of quest romance exemplified by the structuring presence of Cuvier's theory of extinction via geological "revolution" in Kolbert's narrative; Buffon plays a similar role for Claudine Cohen in *The Fate of the Mammoth*. Kolbert notes Cuvier's self-conscious dedication to scientific progress but also his own contemporary reputation as a "poet" (*Sixth Extinction*, 38). Kolbert's source for this designation is Honoré de Balzac, but among writers Lord Byron was the one to exploit this poetic potential most fully in *Cain* and *Don Juan*. Appropriating Cuvier's historiography of revolution for his own mock epic, Byron blithely disregards Cuvier's disclaimer concerning the diminished force of revolutions and envisions a future race excavating the mammoth skeleton of George IV and "other relics of a former world . . . thrown topsy-turvy" underneath "the superstratum which will overlay us."[34] The situation of Byron's future reader here is precocious in the same sense as ours vis-à-vis Buffon's "treasuries of nature": we have an intimate knowledge of the historical causes of the treasuries' depletion that is not available to Buffon's imagined reader of 78,000 C.E. In retrospect, Earth scientists can appear both as the authors of exploded systems and as having an almost vatic power to anticipate Earth system changes.

In light of the resource exhaustion and the destabilized planetary boundaries that have inspired recent projections, the recurring figure of the distant future reader in the science of the Anthropocene must be regarded as purely speculative. The nature of these future "impacts of human activity" makes it extravagant even to speculate that they will remain "observable in the geological stratigraphic record for millions of years into the future," as these accounts routinely assume (Lewis and Maslin, "Defining," 171). Kolbert and Colebrook are among the commentators who do insist on confronting the problem of species survival seemingly bracketed by this quixotic trope of a future geologist, positing a future reader not so much posthuman as inhuman. Kolbert concludes that the "amazing" scope of the current extinction event is "more worth attending to" than "the fate of our own species," even as she recognizes that this conclusion comes "at the risk of sounding anti-human" (*Sixth Extinction*, 268). Borrowing a speculation from Zalasiewicz, she points out that the "biostratigraphic signal" of the Anthropocene will be "permanently inscribed" in the DNA of "surviving . . . stocks" in the future (109)—inscribed and perhaps even legible to the "large, naked rodent, living in caves," which will occupy the niche left vacant by humans (106).[35] Lest we reserve the activity of reading exclusively to our own species, Cary Wolfe points out that language itself, like other human faculties, derives from "ahuman evolutionary processes" and "recursive co-ontogenies."[36]

Focusing on the post-Anthropocene future rather than the evolutionary past, Colebrook envisions an "inhuman perception" for which "the earth, after humans, will offer 'a reading' of a species' history" (*Death of the PostHuman*, 1:24). This type of reading is geological, for Colebrook, because it operates on the "hybrid assemblage of marks . . . that comprises any archive" (34). This inhuman reading practice is also a method of "reading like geology" because geology too reads on a scale that defies narrative, as when Colin Waters and his colleagues refer to technofossils as "the greatest expansion of new metals since the Great Oxygenation Event" of 2.4 billion years ago.[37] Colebrook observes that "the anthropocene thought experiment . . . alters the modality of geological reading, not just to refer to the past as it is for us, but also to *our* present as it will be without us" (28). The future reader, though, already inheres in the modality of "geological reading" as a figure (typically, but not necessarily, in human form) associated with the origins of the geological time scale via theories of the Earth and other forms of scientific romance.

Anticipating inhuman perception is easier said than done. It could be argued that *The Time Machine* (1895), one of Wells's great scientific romances, refers to "*our* present as it will be without us" by way of the Time Traveller himself, who discovers Victorian inventions in a defunct (though suspiciously durable) museum, the Palace of Green Porcelain, some eight hundred millennia in the future. The decaying books in this museum, "though every semblance of print had left them," remind the narrator of his own papers in the *Philosophical Transactions*, a piece of presentism that shows clearly how geological and historical time are conflated in this scenario.[38] Colebrook similarly conflates the two temporalities in her examples from the films of Danny Boyle. Although it is compelling to think of scenes like "the Sydney opera house as a frozen wasteland" as produced by a "geological eye" (*Death of the PostHuman*, 26), they are still apocalyptic scenes and difficult to disentangle from comparable images (such as the beached Statue of Liberty in *Planet of the Apes* or Anna Laetitia Barbauld's ruined London) that convey the romance of extinction.[39] Colebrook also points out an anxiety over the extinction of humanness independent of the fate of the species and mounts a critique of the new vitalism in theory as a refusal to theorize extinction (36–40).[40] Since its inception, the romance of extinction has been defined against the quest for species survival, extending from Buffon's long carbon-based future to Lewis and Maslin's construction of the human species as a "geological superpower."[41] This figure is meant to reinforce their substantive claim that the Anthropocene is an "anti-Copernican revolution" ("Defining," 177), drawing attention to one species' disruptive force, but inevitably the figure also lends weight to the "history of the species" that Dipesh Chakrabarty, among others, has identified as a characteristic narrative form of the Anthropocene.[42]

From Forster through Kolbert, readers of Buffon's *Epochs of Nature* have noted its conjectural elements—speculations derived from conjectural history, the Enlightenment discourse on the history of the species that allowed him to mediate between the human past and the genuinely geological time scale that he envisioned. Forster took issue with Buffon's radical notion that human beings were a late arrival on the scene of antiquity. More like a writer of romances than a naturalist, in Forster's view, Buffon defines the limits of the primeval ocean, produces now-extinct megafauna, and then causes the continents to separate before allowing humans to exist. Forster comments disapprovingly: "At last human beings too become inhabitants of this earth" ("Dr. Forster," 148). Forster's close acquaintance with the ancient historians, and especially his fieldwork among Pacific peoples during three years on board the *Resolution* (1772–75), made him suspicious of theories of human origins that exceeded empirical limits. Cuvier, writing a generation later, ridiculed the absolute dating of formations and species by Buffon and other naturalists who "pile up thousands of centuries with a stroke of the pen," but he accepted the recent arrival of "man, to whom has been accorded only an instant on earth," as a necessary consequence of the record of past extinctions.[43] Buffon's seventh and final epoch, "in which Man assisted the operations of Nature," anticipates some of the problematic aspects of the Anthropocene, particularly the encroachment of a ubiquitous human actor on the stage of geohistory. At the same time, his earlier epochs decenter the human perspective. Although (pace Forster) Buffon raises the possibility of a human presence before the separation of the continents (epoch sixth), this was nevertheless the time of the mammoths and the question of human presence is peripheral to it. Buffon's Age of Man, in other words, implies earlier ages in which humans were not the main actors.

Kolbert's extinction narrative integrates these historical models with current research on extinction as well as human origins, raising the stakes of the threat while also preserving some of the uncertainty surrounding the history of our species in relation to others. Recognizing that Buffon's theory of extinction allowed the human species to appear on the stage of geohistory, she exaggerates his caution considerably to produce a romance of discovery: "with great trepidation, Buffon allowed that this last species—'the largest of them all'—seemed to have disappeared. It was, he proposed, the only land animal ever to have done so" (*Sixth Extinction*, 27). There is an implied contrast between Buffon's hunch and the confidence that Cuvier is able to build as he adds species after species to the roll of extinction after Buffon's *incognitum* (the mastodon, or "Ohio animal").[44] Cuvier accordingly plays a more substantial role in Kolbert's narrative, which returns repeatedly to his thesis of "revolutions on the surface of the earth" (*Sixth Extinction*, 226, 265). Her metanarrative of "the history of the science of extinction" (93) touches on several other episodes that bear on the history of scientific romance, including the controversy over Walter Alvarez's end-Cretaceous asteroid-impact hypothesis in the 1980s.

Kolbert also contributes in her own right to the "unnatural" history of "the species" (the very phrase conveys exceptionalism in somewhat the same manner as eighteenth-century platitudes about "the sex"). She depicts humans as evolutionary "grifters" (*Sixth Extinction*, 202, 204), as the last actors on the evolutionary stage (237), and as the heroic possessors of a "Faustian gene" (251), an "aesthetic mutation" (257) that inspired us—unlike the Neanderthals and other parahuman species that have proliferated in recent years—to cross the oceans. Kolbert shows considerable sympathy toward the Neanderthals, however, and the "cave man"—whether conjectural, as in Buffon, or well buttressed by DNA and fossil evidence, as in Daniel Lord Smail's "deep history"—remains a variety essential for backdating the species as a whole, thereby expanding its geological footprint in the face of extinction. Human origins might also be cited as a characteristic concern of Anthropocene fiction.[45] The particular appeal of Big Bone Lick on the Ohio River, where Kolbert sets her mastodon chapter, is that the sheer quantity of remains there gave a certain symbolic weight to the thesis of extinction long before it was generally accepted. The history of extinctions is therefore a multispecies romance, predicated to some extent on the sympathetic identification promoted by the proliferation of agonies evoking mass extinction. Cuvier's broad analogy between the ruins of antiquity and the fossil record, which he describes as "the ruins of the great Herculaneum overwhelmed by the ocean," makes this identification more explicit.[46]

For Buffon, as for contemporary writers, the prospect of extinction evokes the evolutionary past. Reversion to type, or a kind of regression sparked by evolutionary memory, appears in some contexts as an alternative to extinction. The nostalgia for all things Paleo that pervades U.S. popular culture cultivates a vision of species survival in the face of climate change, as I have argued elsewhere.[47] For Buffon, the fate of the mastodon is bound up with the fate of human giants, an ancestral race inhabiting the border zone between geological and historical time. In one of his numerous variations on the romance of human origins, the last survivors of a giant hominin "nation" migrate across what is now the Bering Strait at roughly the same time as the giant "elephants"; while the mastodons ultimately perish in Central America because they are unable to cross the mountains of Panama, the human giants press on all the way to Patagonia (*Époques*, 193). The appeal for sympathy here is more muted, but this story of the New World elephants' demise arguably makes the survival of the modern Old World elephants more poignant. Since Buffon believes that "the human species is as ancient as that of the elephants," we are warranted in staging this parting as a tearful one, a farewell perhaps even to the "liberty and tranquillity" enjoyed by both species (193, 14) in the warm and humid fifth epoch. Thus a naturalist deeply committed to the domestication of nature, and justifiably associated with European colonialism, swerves from anthropocentrism long enough to contemplate human belatedness and extinction in the guise of nostalgia for a lost Viconian multispecies golden age. Elsewhere in *Époques*, Buffon posits an even earlier evolutionary memory of the upheavals of a younger, hotter planet, with the associated possibility of a reversion to superstitious terror at the sight of earthquakes and volcanoes (206).

The "elephant" standing for the "unknown animal" or mastodon is arguably the main actor of Buffon's *Epochs of Nature*. Buffon himself appears in *The Sixth Extinction* (as do Thomas Jefferson and Cuvier) as an actor in the chapter on the mastodon. Kolbert also develops a trope of "interspecies recognition" (*Sixth Extinction*, 224) that is indebted to Buffon's sympathetic account of the giant "elephant" and echoed in the cover design of her book, which features an upended mammoth skeleton. In Claudine Cohen's reading, "it is not too much to claim that the question of the Siberian mammoth and the 'unknown Ohio animal' is the keystone—maybe the key—of this masterwork."[48] The mammoth also figures prominently in the first, heavily abridged, English translation of *Epochs of Nature* (1785) by William Smellie.

Even genetic engineering cannot erase the evolutionary memory inherited from this extinct ancestor of the megodont, a new draft animal deployed in the postcarbon future imagined by the novelist Paolo Bacigalupi in *The Windup Girl* (2009). The factory that provides one of this novel's main settings is destroyed when panic strikes these animals, and they revert to type: "The sweet stink of human offal permeates the air. Gut streamers decorate the megodont's circuit around its spindle. The animal rises again, a mountain of genetically engineered muscle, fighting against the last of its bonds. . . . Another megodont rises to its hind legs, trumpeting sympathy."[49] While Bacigalupi's human characters are still very much historical actors, he situates them in deep time through the laboratory-enhanced atavism of other species and through catastrophic climate change. His vision of megodont-driven factories reveals this cli-fi novel's affiliation with the scientific romance tradition.

The fossil fuel subtext lurks just beneath the multispecies romance. In the absence of fossil fuels, Bacigalupi's industrialists harness the energy genetically stored in extinct animal life. In Buffon's thermal-maximum epochs, the "primitive vigor" suggested by the size of fossil bones is confirmed by the record of teeming early plant life deeply inscribed in the coal measures. Smellie only included an abridged version of *Epochs* in his nine-volume translation of Buffon's *Natural History* because he felt that this "theory" was "perhaps too fanciful to receive the general approbation of the cool and deliberate Briton."[50] Yet the two substantive parts of *Epochs* that survived his abridgment unscathed are among the most fanciful of all. Smellie includes Buffon's seventh epoch, his romance of inexhaustible coal, in its entirety, along with all his excursus notes, on the grounds that these contain "interesting facts" that are "too important to be omitted." These notes contain traces of the synchronic multispecies narrative that Buffon suppressed in revising the main text shortly before publication, as documented in Jacques Roger's critical edition of the work; three of the longest notes are dedicated to a theory of gigantism encompassing the extinct "elephants," the giant ammonites, and tropical fossil flora associated with the coal measures. As Smellie renders it, "Nature was then in her primitive vigour. The internal heat of the earth bestowed on its productions all the vigour and magnitude of which they were susceptible. The first ages produced giants of every kind" ("Facts," 303). Similarly, the deleted passage on the common age of giant elephants and giant men correlates these animal species with the fossil plants that function as "monuments" of a planet covered with forests like those of modern-day Guyana (*Époques*, 159n).

These fossil plants, like the mastodon molars engraved to illustrate *Epochs of Nature*, serve as monuments of past events, but unlike the fossil teeth they have a double function. In their superabundance, they suggest a permanent endowment, an archive conserving the gigantic energy of the young Earth. When Buffon discusses the formation of coal in the third epoch, he emphasizes the "fecundity" of the younger planet that produced both plants and animals adapted to much higher temperatures; the excursus note cites numerous analyses of fossil impressions from coal mines that seem to describe fish and plants that "do not belong to living species" (*Époques*, 78). His revisions show that he did not anticipate controversy over extinction so much as over the issue of human exceptionalism, which his long time scale seemed to threaten. Accordingly, the seventh epoch, written last, returns to the subject of coal at length to show that the superabundance of the third epoch persists as a guarantee that human intervention will permanently transform nonhuman species and render them "useful."[51] Thanks to fossil fuels, the intervention extends to climate engineering, for these vast "treasuries" make it downright easy to "heat the earth" (*Époques*, 215).

Cuvier argues, similarly, that "nature seems to have put [coal] in reserve for the present age." Commenting on Cuvier's sentence, Martin Rudwick terms it "a secular version . . . of the view commonly expressed in Britain at this time, that the ancient coal deposits had been stored for eventual human use, by the care of divine providence."[52] John Whitehurst, writing like Buffon in 1778, spoke somewhat more cautiously of fossil plants "not known to exist in any part of the world in a living state" (*Inquiry*, 203). These impressions, however, combined with those of recognizable tropical plants, provide "a certain indication of coal" in the underlying strata. In Whitehurst's idiom of natural theology, the certainties inscribed in the order of the strata correspond more closely to the "book of nature," written for a future reader who would grow more literate as the value of coal increased. The formation of the Earth speaks clearly of coal and other mineral resources in Whitehurst's reading, indicating (like Buffon's) the importance of fossil fuels as a driver of Enlightenment geochronology. In this way, the human species actor and its dependence on fossil species is already implicit in the GTS.

Giants and their inexhaustible treasures are certainly the stuff of romance. The tropes of extinction and the future reader are less readily recognizable, but all depend on scale. As Benjamin Morgan argues in this volume, the "disjuncture" between human and nonhuman scales is itself a problem of form. The events of species extinction and the eventual fossilization of the present both transpire on a scale that is ultimately incompatible with that of human generations, while the scalar disjuncture between the giant and the romance hero, if somewhat less extreme, makes the giant's defeat all the more satisfying for its sheer improbability. The giant species that populate Buffon's fossil record stand in somewhat the same relation to the larger problem of the geological time scale. "Our too brief existence," he declares, constrains us to minute analysis of the "numerous centuries required to produce the mollusks with which the earth is filled; [of] the even greater number of centuries that have passed since their shells were transported and deposited," and so on through the processes of petrifaction and erosion (*Époques*, 41). Similarly, the gradually decreasing size of these specimens marks the stages of an otherwise unintelligible history. Scaling up or down seemingly at will, Buffon makes the point that the history of other species is indispensable for arriving at a conception of geological time that is at least commensurate with "the limited power of our intelligence" (40n)—but may still be far short of its true extent. Buffon's assurances of an inexhaustible supply of coal, juxtaposed against the few centuries it took to exhaust these reserves (or nearly so), reproduces the disjuncture between geological and historical time: the portion of the fossil record that inspired Buffon's long geological time scale, and secured the place of the human species within it, was rendered historical by its extraction.

The scientific romance of geological time depends on figures that arise from the problem of scale, performing a work of mediation that necessarily remains incomplete. The Anthropocene puts new pressure on these figures. Clive Hamilton rightly notes that the very notion of humans changing the Earth system challenges the basis of the GTS ("Getting," 106). Hamilton and Grinevald define the novelty of the Anthropocene through this disjuncture, insisting that it cannot be construed as happening "incrementally over deep time" ("Was the Anthropocene," 66). At the same time, this idea of a rupture is a counterpart of the time parallax invoked by earlier commentators pointing out that changes such as extinction (Kolbert) or global cooling (Buffon) are imperceptibly slow from the perspective of human time. Mobilizing human actors in the space of deep time has been a major concern of geological romances since this science began to be formalized. Locating human origins in this space allows a firm boundary to be drawn between geological and historical time, and the anxiety over this demarcation point that we see in Buffon's numerous variations on human prehistory is still at work in Lewis and Maslin's pursuit of a global stratotype section and point, or golden spike, to locate the exact onset of the Anthropocene. The difference between the two also reveals that "human time" itself is far from homogeneous. The romance form historically offers many possibilities for the disruption of "homogeneous, empty time," and since the inception of the GTS it has proven adaptable to the challenge of negotiating multiple discrepant temporalities.[53]

The formal continuities between Buffon's scientific romance and Anthropocene popular science indicate that a revival is under way, a revival soon to be consolidated by the new translation of *Epochs of Nature* by Zalasiewicz and his colleagues. They justify this new translation by arguing that "the sciences have come full circle," both in the sense that natural history is reunited with cosmology and in the sense that the process of scientific specialization has reached a limit: "it has become increasingly clear that one has to understand not just the parts (in minute detail) of the whole, but also the 'whole' itself."[54] The scientific romance flourished as a narrative nonfiction form in an era of intensified resource extraction and now offers new possibilities both for a critique of our romance with fossil fuels and for an imaginative reckoning with resource exhaustion.

## Notes

1. Jacques-Henri Meister, "Avril," *Correspondance Littéraire* 10 (1779): 169–72, 169.

2. Clive Hamilton and Jacques Grinevald, "Was the Anthropocene Anticipated?," *Anthropocene Review* 2.1 (2015): 59–72 (hereafter cited parenthetically in text).

3. Zalasiewicz has generously provided me with the draft of the introduction to the new translation of Buffon. Page references throughout this chapter are to the manuscript copy in my possession. See Georges-Louis Leclerc, le Comte de Buffon, *The Epochs of Nature*, translated, edited, and compiled by

Jan Zalasiewicz, Anne-Sophie Milon, and Mateusz Zalasiewicz (Chicago: University of Chicago Press, 2018).

4. It did matter increasingly to readers of Buffon in the late eighteenth century and into the nineteenth, as Wolf Lepenies has shown in *Autoren und Wissenschaftler im 18. Jahrhundert* (Munich: Hanser, 1988), 61–90.

5. For these criticisms, see Stanley C. Finney and Lucy E. Edwards, "The 'Anthropocene' Epoch: Scientific Decision or Political Statement?" *GSA Today* 26.3 (2016): 4–10; Whitney J. Autin and John M. Holbrook, "Is the Anthropocene an Issue of Stratigraphy or Pop Culture?" *GSA Today* 22.7 (2012): 60–61. Clive Hamilton's insistence that Lewis and Maslin's work is not proper Earth system science belongs to the same category of objections. See Hamilton, "Getting the Anthropocene So Wrong," *Anthropocene Review* 2.2 (2015): 102–7 (hereafter cited parenthetically in text).

6. Among the traditional topoi that inform this aspect of geological romance, the prophecies of King Arthur's return in Arthurian romance stand out for implying a future reader. The mode of cyclical time in romance, associated with its roots in oral tradition, as instanced in the Green Knight's return in *Sir Gawain*, is also relevant here. Chaucer's envoi to *Troilus and Criseyde* ("Go, litel bok") addresses a future reader, though not a far-distant one. In geology proper, see especially the cartoon *Awful Changes* (1830) by Henry De La Beche and the discussion of it in Martin Rudwick, *Scenes from Deep Time* (Chicago: University of Chicago Press, 1992), 48–49.

7. Martin Rudwick, ed. and trans., *Georges Cuvier, Fossil Bones, and Geological Catastrophes: New Translations and Interpretations of the Primary Texts* (Chicago: University of Chicago Press, 1998), 183.

8. Elizabeth Kolbert, *The Sixth Extinction: An Unnatural History* (New York: Holt, 2014), 105 (hereafter cited parenthetically in text).

9. Claire Colebrook, *Death of the Post-Human: Essays on Extinction* (Ann Arbor: Open Humanities, 2015), 1:39 (hereafter cited parenthetically in text). http://www.openhumanitiespress.org/books/titles/death-of-the-posthuman.

10. Buffon, *Les Époques de la Nature: Édition Critique*, ed. Jacques Roger (1962; repr., Paris: Éditions du Muséum, 1988), 40 (my translation) (hereafter cited parenthetically in text).

11. Zalasiewicz et al., unpublished introduction to *Epochs of Nature*, 26; Hamilton, "Getting," 2 ("standard-bearers"). On deep time and epic, see Mark McGurl, "The Posthuman Comedy," *Critical Inquiry* 38.3 (2012): 533–53; and Wai Chee Dimock, "Low Epic," *Critical Inquiry* 39.3 (2013): 614–31.

12. Paul Harvey, comp., *The Oxford Companion to Classical Literature* (1937; repr., Oxford: Oxford University Press, 1984), 288; B. P. Reardon, *The Form of Greek Romance* (London: Routledge, 2014), 129; and John Dean, "*The Odyssey* as Romance," *College Literature* 3 (1976): 228–36, 229. See further Steve Mentz, *Romance for Sale in Early Modern England: The Rise of Prose Fiction* (Aldershot: Ashgate, 2006), 73–75.

13. See Holmes, *The Age of Wonder: How the Romantic Generation Discovered the Beauty and Terror of Science* (New York: HarperCollins, 2008).

14. Simon Lewis and Mark Maslin, "Anthropocene: Earth System, Geological, Philosophical, Political, and Paradigm Shifts," *Anthropocene Review* 2 (2015): 108–16, 112.

15. Wells, "Preface to *The Scientific Romances*," in *H. G. Wells's Literary Criticism*, ed. Patrick Parrinder and Robert M. Philmus (Totowa: Barnes and Noble, 1980), 240–45, 242. Also see Brian Stableford, *Scientific Romance in Britain, 1890–1950* (New York: St. Martin's, 1985), esp. chap. 2.

16. Zalasiewicz et al., unpublished introduction, 25.

17. Fredric Jameson, *The Political Unconscious* (Ithaca: Cornell University Press, 1981), 81.

18. Stephen Best and Sharon Marcus, "Surface Reading: An Introduction," *Representations* 108.1 (2009): 1–21.

19. Elaine Freedgood and Cannon Schmitt, "Denotatively, Technically, Literally," *Representations* 125 (2014): 1–14.

20. Simon L. Lewis and Mark A. Maslin, "Defining the Anthropocene," *Nature* 519

(2015): 171–80, 176–77 (hereafter cited parenthetically in text).

21. Falko Schmieder, "Urgeschichte der Nachmoderne: Zur Archäologie des Anthropozäns," *Forum Interdisziplinäre Begriffsgeschichte* 3 (2014): 43–48 (my translation).

22. Roger transcribes Guettard's manuscript letter in its entirety in the notes to his critical edition of Buffon (*Époques*, cxxxix–xl, n. 7).

23. "Dr. Forster an Prof. Lichtenberg," *Göttingisches Magazin der Wissenschaft und Litteratur* 1 (1780): 140–57 (my translation) (hereafter cited parenthetically in text).

24. Playfair qtd. in Roy Porter, *The Making of Geology* (Cambridge: Cambridge University Press, 1977), 1. On Playfair and the romance of disciplinary progress, see also Noah Heringman, *Romantic Rocks, Aesthetic Geology* (Ithaca: Cornell University Press, 2004), 270–72.

25. Zalasiewicz et al., unpublished introduction, 9, 23, 17.

26. Ruddiman, "The Anthropogenic Greenhouse Era Began Thousands of Years Ago" (2003), qtd. in Hamilton and Grinevald, "Was the Anthropocene," 62. See also Finney and Edwards, "The 'Anthropocene' Epoch."

27. Zalasiewicz et al., "Colonization of the Americas, 'Little Ice Age' Climate, and Bomb-Produced Carbon: Their Role in Defining the Anthropocene," *Anthropocene Review* 2.2 (2015): 117–27, 124.

28. Zalasiewicz et al., unpublished introduction, 3.

29. John Whitehurst, *Inquiry into the Original State and Formation of the Earth*, 2nd ed. (1786; repr., New York: Arno, 1978), 257 (hereafter cited parenthetically in text). The first edition of the *Inquiry* was published in 1778.

30. Dipesh Chakrabarty points out the timeliness of "species thinking" in "The Climate of History: Four Theses," *Critical Inquiry* 35.2 (2009): 197–222; cf. Fredrik Albritton Jonsson, "The Industrial Revolution in the Anthropocene," *Journal of Modern History* 84 (2012): 679–96. One of the most compelling historians of "the" species is (in my view) Daniel Lord Smail; see *Deep History and the Brain* (Berkeley: University of California Press, 2008) and subsequent works.

31. "To use up" is a slight exaggeration here, since fracking and other new invasive methods promise to sustain the supply of fossil fuels temporarily. Meanwhile, carbon emissions pose the more critical problem. On resource exhaustion, see Jonsson, "Industrial Revolution."

32. Cuvier qtd. in Kolbert, *Sixth Extinction*, 43.

33. Playfair qtd. in Porter, *Making*, 1; *The Mary Shelley Reader*, ed. Betty T. Bennett and Charles E. Robinson (Oxford: Oxford University Press, 1990), 35.

34. Byron, *Don Juan*, 9.37–39. See also Ralph O'Connor, "Mammoths and Maggots: Byron and the Geology of Cuvier," *Romanticism* 5 (1999): 26–42.

35. There is perhaps an inhuman aspect as well to Freedgood and Schmitt's repeated injunction to "read more!" ("Denotatively," 9), which estranges the reader, or readerly capacity, as much as the text itself (12).

36. Cary Wolfe, *What Is Posthumanism?* (Minneapolis: University of Minnesota Press, 2010), xxii.

37. Colin N. Waters et al., "The Anthropocene Is Functionally and Stratigraphically Distinct from the Holocene," *Science* 351 (January 2016): 137–52, 139. On technofossils, see also Jussi Parikka, *A Geology of Media* (Minneapolis: University of Minnesota Press, 2015), chap. 5.

38. H. G. Wells, *The Time Machine* (New York: Bantam Classics, 2003), 83.

39. Zoe Crossland raises the dystopian sublime as a political problem by framing the archaeology of the Anthropocene as a project of turning "our gaze away from [this] projected dystopia ... to the present and past conditions that underwrite its potential unfurling." Crossland, "Anthropocene: Locating Agency, Imagining the Future," *Journal of Contemporary Archaeology* 1.1 (2014): 123–28, 127.

40. Thou shalt believe in Badiou, Harman, Brassier; / Thou shalt not set up Arendt, Butler, Negri. See Byron, *Don Juan*, I.ccv.

41. Lewis and Maslin, "Anthropocene," 112.

42. Chakrabarty, "Climate of History."

43. Cuvier, "Discours preliminaire," *Recherches sur les Ossemens Fossiles de Quadrupèdes* (1812), in Rudwick, *Georges Cuvier*, 228, 252.

44. Buffon himself, though, surmises that many of the largest creatures found in the fossil record might now be extinct (e.g., the giant ammonites produced by Earth's "primitive vigor"). And certainly he is far from timid in his assertions concerning extinction. (Kolbert may be thinking of the conflict he anticipated with the Sorbonne over his time scale.)

45. Kate Marshall traces a kind of atavistic "production of art on Pleistocene desert sands" in the Anthropocene novel in her "What Are the Novels of the Anthropocene? American Fiction in Geological Time," *American Literary History* 27.3 (2015): 523–38, 524. Max Frisch's *Der Mensch erscheint im Holozän*, translated as *Man in the Holocene* (1979; New York: Harcourt Brace Jovanovich, 1980), might also be cited as relevant here.

46. Georges Cuvier, *Essay on the Theory of the Earth*, 4th ed., ed. and trans. Robert Jameson (Edinburgh: Blackwood, 1822), i.

47. Noah Heringman, "Evolutionary Nostalgia in the Anthropocene," *minnesota review* 83 (2014): 143–52.

48. Claudine Cohen, *The Fate of the Mammoth: Fossils, Myth, and History*, trans. William Rodarmor (Chicago: University of Chicago Press, 2002), 96.

49. Paolo Bacigalupi, *The Windup Girl* (San Francisco: Nightshade, 2009), 16. The Windup Girl herself, Emiko, also reverts to type in the course of the novel.

50. "Facts and Arguments in Support of the Count de Buffon's *Epochs of Nature*," in Buffon, *Natural History, General and Particular*, 2nd ed., trans. William Smellie (London, 1785), 9:258–410, 258 (hereafter cited parenthetically in text).

51. Ironically, Buffon's attempt to make his system *more* orthodox by postdating the species backfired, and it was this portion of the text that the Sorbonne condemned.

52. Rudwick, *Georges Cuvier*, 124–25, 125n34.

53. On homogeneous, empty time, see Walter Benjamin, "Theses on the Philosophy of History," in his *Illuminations* (New York: Schocken), 261.

54. Zalasiewicz et al., unpublished introduction, 1. The editors also credit Buffon as the "founder" of the "geochronological time scale" (14).

# 4.

# Punctuating History Circa 1800
## The Air of *Jane Eyre*

*Thomas H. Ford*

Considered in Anthropocene terms, the end of the eighteenth century and the beginning of the nineteenth mark the onset of carboniferous industrialization. Seen in more traditionally literary historical terms, these are dates central to Romanticism. I take the relationship between these two sets of terms to be one of historical "synomonymy," a neologism that I discuss in more detail below, but which can be understood for now as naming a semantic event of simultaneous synonymization and desynonymization. Romanticism's chronological coincidence with the Anthropocene is not just a coincidence. Underlying the contemporaneity of these categories is a matrix of shared conceptual presuppositions and unresolved philosophical and social questions; both are inspired by new capacities to channel planetary forces to productive ends and by the consequent acceleration of intertwined ecological and historical changes at global scales. If not entirely synonymous, the categories are then at least semantically convergent. On the other hand, however much our reading practices may still be conditioned by aesthetic, institutional, and even ecological forms that endure from this moment, the Anthropocene and Romanticism must also be understood as noncoincident, conceptually as well as historically. More important here than the plurality of suggested alternative Anthropocene dates is the historical fact that the Romantics could name themselves but not the Anthropocene, for the concept was simply unavailable to them. The ecological changes to which the category refers had mostly not yet happened; the technosocial drivers of those changes were still only incipient; the forms of knowledge required to conceptualize it were unimagined. Our Anthropocene past was their non-Anthropocene future.

Reading Romantic texts in the Anthropocene is thus a paradoxical practice. On the one hand, Romantic artworks are legible as Anthropocene artworks at an unquestionably material level. But on the other, that legibility is never more than metaphoric, consisting in indirectly allusive anticipations that come to light only through speculative and retrospective interpretations of past literary artworks as having written the distant Anthropocene present. My example for this cross-hatching of material and metaphoric links between past and present is Charlotte Brontë's 1847 novel *Jane Eyre*, which I read as a text dedicated to remediating the fading cultural potentiality of its recent literary past into a changed future. Even more than its explicit intertextual references—the second volume of Thomas Bewick's *A History of British Birds* (1804); Byron's *The Corsair* (1814); Walter Scott's *Marmion* (1808)—or its allusive verbal echoes of such texts as Mary Wollstonecraft's *A Vindication of the Rights of Woman* (1792) and Percy Shelley's *A Defence of Poetry* (1821), it is *Jane Eyre*'s formal reworking of atmosphere that allows it to be read as at once a late Romantic text and a work that anticipates the Anthropocene: as a novel, that is, which looks back to look forward. But before considering how *Jane Eyre* revisits Romanticism to punctuate atmospheric history anew, let me first suggest in what sense we might speak of the Anthropocene and Romanticism as being historical synonyms circa 1800, and why in consequence the Anthropocene Earth system can be read as a literary artwork.

Only around 1800 was literature first defined in Europe as a specialized domain of cultural practice and scholarly research in its own right.[1] The basic conceptual and institutional architecture of modern literary studies emerged as part of Romanticism's comprehensive reconfiguration of the categories of cultural description and evaluation. One vital impetus for this new disciplinary formation was that literature was seen to provide an answer to a specific philosophical problem. The problem had been posed most starkly in Immanuel Kant's critical divisions between things as they are in themselves and things as we know them, between material necessity and moral agency, and so between object and subject. In the Kantian aftermath, this became known as the problem of "the absolute" or "the unconditioned": how to think of mind and nature as independent and irreducible to one another, but also as ultimately unified, continuous with one another.[2] The aim was to reconcile conceptual thought with its bounded, physical, nonconceptual embodiment so as to arrive at "intellectual intuition," in which subject merges with object.[3] With Romanticism, the route to the achievement of this "absolute" was seen to run not via philosophy but through art, and paradigmatically through literature. That claim is still central to literary studies, inasmuch as it still underwrites many of the definitions of literature that organize the discipline.[4]

Hegel summarized this position in the opening pages of his lectures on aesthetics: modernity produces an "opposition in man which makes him an amphibious animal, because he has to live in two worlds which contradict each other."[5] On the one hand, humans are natural, embodied objects in the world, subject to the same determining principles that govern all other material phenomena. On the other hand, humans are self-defining and reflective rational agents, freely constructing the cultural and ethical norms required for actions to be understood as bearing meaning. In the philosophy of German Romanticism, art presented the means "to dissolve and reduce to unity" this fundamental contradiction "between the abstractly self-concentrated spirit and nature" (Hegel, *Aesthetics*, 56). Art was seen as capable of mediating between the otherwise irreconcilable opposites of historical intelligibility and the mute material world, and so could resolve the unresponsive otherness of nature with communicative thought. Art, in other words, was what made sense of the world for modern philosophy.

Literature—and lyric poetry in particular—was thought to provide an especially telling instance of this aesthetic presentation of the absolute. This was because in literature the paradoxical demand to think objectively and subjectively at the same time received its most acute formulation. The lyric poem "explores the most subjective, nonconceptual, and ephemeral phenomena," and yet it does this in language, the medium of objective, conceptual knowledge.[6] In Romanticism, the artwork was understood to be irreducibly material, bound up inextricably with the contingencies of its empirical appearance in a certain time and place. Romantic lyric poetry sought to put this incommunicable singularity into words, making it public and even potentially universal. For Friedrich Schlegel, the philosophical importance of poetry was then that it showed "how the human spirit impresses its law on all things and how the world is its work of art."[7] In Romanticism, literature's emergence as a modern aesthetic category rested on this claim that it resolved the philosophical division of mind from nature, of the self-assertions of the human spirit from the lifeless necessity of the object world.

The Anthropocene concept reworks these divisions. It registers the increasingly complex historical entanglement of human actions with natural appearances, and so also echoes Romantic formulations that attributed to literary artworks a similarly mediating ontological position. This construction of the literary artwork as a fraught synthesis of human and nonhuman orders of existence was a central element of German aesthetic thought following Kant, and in comparable ways of British literary and critical culture in the same period. As analyzed by Kant in the third *Critique*, apprehending an artwork as beautiful requires us to view it as if it were natural. We must see it as a found object, not as something made. Meanwhile, for Kant, apprehending natural beauty requires us to view nature as if it were an art object—as made, not found.[8] According to him, beauty appears through this crossing of the categories of art and nature, slipping between human technology and natural appearance. Around 1800, the object of an aesthetic judgment was accorded a dual ontology, neither wholly assimilable to human systems of meaning nor ever seamlessly continuous with the natural world. This doubled ontology underwrote assertions of the autonomy of the aesthetic, and so was the basis of what was seen to follow from that autonomy: the artwork's infinite interpretability, its exteriority to social purposes, the quasi-logical necessity of its construction, the potential transhistorical public it invokes, and so on.

The structure of the Romantic experience of beauty positioned its object as being neither wholly natural nor purposefully made, but rather somehow both at once. That object was nonconceptual, a particular matter of sensation, but it was also semiotically communicative and thus conceptual. It appeared to suspend or even efface the distinction between symbolic actions and natural appearances, and it did this even while that distinction remained a condition of its intelligibility, for the very content of judging something to be beautiful was in effect nothing but the infinite shuttling back and forth across this distinction, referring each side to its opposite. The beautiful object was shaped and experienced in human time, but also seemed to reach out to encompass nonhuman temporalities; as such, it addressed a not-yet-existent, generalized, collective humanity, a universal horizon of human spirit. Our contemporary Anthropocene condition finds an uncanny fit with this Romantic model of aesthetic experience. The Anthropocene world speculatively appears as a planetary literary artwork, as Schlegel anticipated, insofar as it is a world that demands to be seen as expressive and meaningful, not despite but because it is a world its Romantic makers never intended.

Sianne Ngai has called attention to what she calls the "challenging doubleness" of aesthetic concepts; these are mid-range concepts, she writes, that provide "ways of speaking or aspects of human intersubjectivity [that] routinely intersect with qualities or aspects of the thing world."[9] Understanding aesthetic categories to be at once "subjective and objective, evaluative and descriptive, conceptual and perceptual," Ngai shows that this liminal status derives from the paradoxical character of aesthetic judgment in and after Kant.[10] To describe an object as beautiful involves treating what seems to be a subjective judgment as a statement of objective fact, collapsing the performative "I deem this to be beautiful" into the constative "this is beautiful." In line with her interest in commodification and mass culture, Ngai focuses on "minor" aesthetic categories. But it is the canonical categories of high Romantic aesthetics that might be more critical for readers in the Anthropocene, inasmuch as it was with these categories that the semantic doubleness identified by Ngai framed powerful claims about the philosophical significance of aesthetic experience.

As Ngai shows, aesthetic categories toggle between subjective judgment and objective evaluation, two distinct classes of speech act. In categories such as the beautiful, the sublime, and the picturesque, that semantic duality was used to assert art's capacity to reconcile subject with object, and human freedom with material determination. The beautiful typically presented a spontaneous harmony between the human mind and the natural world. The sublime conversely staged their rift, opposition, or even agonistic struggle. The picturesque occupied an uncertain middle ground, acculturating the natural world by leaving "unscenic nature" out of the frame, but also naturalizing culture, occluding "the dark side of the landscape"—the human labor that created that landscape, and the social relations that governed it.[11] These three categories named distinct aesthetic logics for redescribing relationships between social meaning and nonhuman being, and between technological artifice and natural processes. They offered diverse answers to philosophical problems of how to mark and value nature, and of how to coordinate and measure contingency against design, singular material embodiment against conceptual generality. And this same linguistic ambiguity, identified by Ngai in Romantic models of aesthetic judgment, now also characterizes the Anthropocene concept.

In an analysis of the rhetorical work performed by the term, Noel Castree has shown that the Anthropocene is likewise poised between objective report and constructive interpretation, interlinking semantic structures of description with those of evaluative acts of judgment.[12] Applications of the Anthropocene concept are never just straightforward objective descriptions of facts; they are also judgments designed to traverse the linguistic divides separating scientific knowledge from public understanding. The Anthropocene concept then always involves a mode of address, convening an imaginary community; it has even been taken to imply the redefinition of that coming community as an extended field open to nonhuman actants. The Anthropocene semantically converges in these ways with some key Romantic categories of literary interpretation—not only with beauty, sublimity, and the picturesque but, even more insistently, with the category of literature itself.

To be sure, Romanticism said neither the first nor the last word about beauty or the sublime—or indeed about Romantic literature. In the Romantic period, however, aesthetic categories were not only positioned as conceptual instruments for asserting large-scale philosophical claims about the relationship between mind and nature; they also became means for historicizing that relationship. Historical temporality touched all aesthetic categories but was articulated most systematically in Hegel's grand sequence of artistic logics (Symbolic, Classical, Romantic), which enacted through time a dialectic of subjective self-construction and objective materialization. Hegel's example reminds us that historical periodization—the act of naming a semiautonomous chunk or segment of historical experience, an identifiable and coherent historical atmosphere—shares critical semantic and cognitive features with the act of aesthetic judgment more generally. It reminds us, too, why the notorious ambiguity of Romanticism, which names both a temporal segment and a transhistorical artistic style or mood, is such a central example of how cultural-historical period terms can also function as aesthetic categories. Around 1800, the historical period emerged within literary culture as a category that interrelated the aesthetic legibility in the present of texts from the past with the legibility of the present as history.[13] Such formulations implicitly located an aesthetic dimension within all acts of historical periodization. They called particular attention to those that name the period of the historical present, as now takes place with the Anthropocene concept.

The categories of literature and the literary artwork define a specific art form and its disciplinary study. But other aesthetic categories cut across different art forms: a painting or piece of music may be said to be sublime or Romantic just as a poem may be. A picturesque poem is a poem about a natural landscape framed as an imagined painting, or seen through a colored lens or reversed in a convex black mirror—as circumscribed, in other words, by an intermedial array of aesthetic practices, technologies, and theories.[14] But it is specifically literary acts of aesthetic categorization that provide most purchase on the paradoxes of Anthropocene historicism, because with literature the paradoxical construction of the Romantic artwork—at once singular and general, nonconceptual and conceptual—achieved its most philosophically significant formulations. For philosophy circa 1800, literature said something philosophy could not itself say. The fact that literature said this unsayable thing (rather than depicting it or sculpting it, for instance) meant that literature articulated its nonconceptual content in the medium of philosophy's own concepts: the public and even speculatively universal medium of language. Today, it now also entails that the full meaning of Romanticism's semantic convergence with the Anthropocene must be substantiated through individual acts of reading literary texts. The terms of its historical and philosophical legibility necessarily refer to those of its specifically literary legibility, for it is only in the particularities of literature, at least according to this critical philosophy, that the unsayable and the unintendable arrive at verbal expression. Literature says unsayable things and speaks beyond the limits of strictly human intentions. That might well be why we read it; that is, at any rate, what the word "literature" came to mean around 1800. And if the Anthropocene conforms in important ways to a Romantic theory of the artwork, making it possible to present the Earth system itself as the total Romantic literary work of art—my speculative synonymization here—then in studying literary artworks from this period we might also find delineated otherwise unknowable elements of what the Earth system now means. Turning back to Goethe, Charlotte Smith, Friedrich Hölderlin, John Keats, or William Wordsworth offers a way of interpreting the Anthropocene; in them we discover our unrecognized ecocontemporaries, the poetic witnesses of our geohistorical moment. Again, this is not because the Anthropocene began circa 1800, when these poets were living and writing. Rather, it is because any decision about when the Anthropocene began will move unstably and Romantically between empirical knowledge claims and the reflective self-implication of the speaking subject. And as such, it resonates with a literature that already spoke of a world in which we half create what we perceive.

But that resonance also registers a break. The climate of our history, with atmospheric $CO_2$ exceeding 400 parts per million at the time of writing, is not theirs, when levels were around 280 parts per million. Readers of this book will likely be familiar with the calls, prompted by such Anthropocene concerns, for new kinds of transdisciplinary knowledge that better capture the socioecological dynamics these changing figures index. Readers will be familiar, too, with the way in which the Anthropocene seems to compel the reconception of textuality at a more material level. Ontologically speaking, in the Anthropocene, writing must always be understood as writing on the world, whatever else it may be. Any instance of text, whatever it might say, is also in social and material terms so many kilograms of embodied $CO_2$, a future thickness of submarine limestone, and so is potentially legible climatologically and stratigraphically, as well as in any more traditionally literary sense. That shift entails changes to conceptions of the materiality of literary language, which must now be expanded to incorporate these aerial and stony dimensions. In this respect, the Anthropocene could be said to demetaphorize a long-standing set of atmospheric and lithic figures of poetic language: the poem as a sigh, breeze, or inspiration, from Petrarch's "l'aura" to Shelley's "O wild West Wind," or as an inscription, stonily epitaphic and monumental, more durable than bronze—inscriptions from Horace to Ozymandias. We might see this same shift in the sudden comprehensibility in the Anthropocene of what Kant called the "cipher language" (*Chiffresprache*) of nature. There, in what for Kant was the language of nature's infinitely aesthetic ideas, we can now recognize as being unmistakably spelled out, as if in a collective material unconscious, the disastrous truths of our current form of human life. Nature's language is ciphered no longer because we know it as our own, however estranged. We have then to come to terms with a jarring and newly nonmetaphoric legibility common both to the natural world and to its human representations and interpretations. Reading literary texts from the past can help in this task by reminding us what those metaphors once meant, even as they now disappear into the indefinite distance of an unrecoverable natural climate. I will leave lithic metaphors for another occasion (since stone is another matter) and turn to the air of *Jane Eyre*.

Raymond Williams once described the novels of the Brontës as the continuation of the Romantic lyric project by other means.[15] In Charlotte's case, those means involved the novelization of Romantic lyric atmosphere. "Atmosphere," a word first coined in Latin early in the seventeenth century, remained a term essentially restricted to natural philosophy for most of the next two hundred years.[16] It was a physical concept, encountered in fields such as optics, pneumatics, and chemistry. Only around 1800 did the term take on what we now tend to think of as its metaphoric dimension, as a word synonymous with mood, tone, or attunement, and used to name a particular affective quality or ambience, the specificity of a cultural moment, or indeed the distinctive character, the unique "atmosphere," communicated by a text. Between 1790 and 1815 a host of new "atmosphere of . . ." formulations entered the language: in the philological record we encounter, for the first time, atmospheres of sentiment, philosophy, mind, a party, a court, oppression, corruption, freedom, a playhouse, and a college. We also find new linguistic "atmospheres," which run from "poetic atmosphere" to "typographic atmosphere." Certain related phrases, like "climate of opinion," had appeared somewhat earlier, but these became significantly more prominent around 1800, as both climate and atmosphere underwent a rapid semantic expansion to become the central terms of a new critical vocabulary for describing the sense of a surrounding, formative, cultural world and for conceiving of the present as such a historically specific world. One might speak, with regard to these semantic developments and their innumerable correspondent breezes, of an atmospheric Romanticism.

Today, when atmosphere or climate is used to refer to social, political, cultural, or other human historical forms or processes, those words are routinely taken to be metaphors, broadly speaking. But around 1800, particularly in poetic contexts, this new dimension was not yet so clearly marked as metaphoric. To the contrary, when in Britain Wordsworth, Samuel Taylor Coleridge, Thomas De Quincey, Ann Radcliffe, William Hazlitt, Keats, and Shelley or when in Germany Lichtenberg, Herder, Kant, Schiller, Goethe, Novalis, and Schlegel used these words to refer to an unarticulated totality of imaginative feeling, their use was largely literal and material. They attributed atmosphere to words because they understood words to be shaped bodies of air. "The breath of our mouths," Herder wrote, "is the picture of the world, the type that exhibits our thoughts and feelings to the mind of another. All that man has ever thought, willed, done, or will do . . . upon Earth, has depended on the movement of a breath of air."[17]

Throughout the twentieth century, use of the atmospheric lexicon could be divided relatively cleanly between metaphoric and literal registers. But it would be a mistake to apply that semantic division retrospectively onto writing from this earlier period. Then, "atmosphere" was instead a term closer to what is now sometimes called "natureculture." It was a singular linguistic element that was also taken to be the substrate of language in general, and so hovered blurrily between singularity and generality and between metaphor and material fact. Atmospheric Romanticism was then shaped by two opposing dynamics. On the one hand, it was a field of differentiations; atmospheres specified and diverged. Atmospheric language not only was used widely in this period for definitions of literary and aesthetic autonomy, but also played a comparable role in disciplinary self-definitions of modern natural sciences. On the other hand, atmosphere remained the linguistically common air these disparate projects shared, a lexicon through which they exchanged and communicated elements of meaning. In this sense, atmospheric Romanticism presented a paradoxical event of differentiation and desynonymization as well as of totalization and synonymization. An atmosphere was unique, but it was also internally multiple and compounded; it was particular and could be distinguished from other atmospheres, and yet its boundaries were at best vague, as it drifted indistinguishably into what surrounded it. It was bound to the here and now, but it was also mobile and citable. Atmospheric language specified, but it also intermixed; it determined but also rendered indeterminate.

"A waft of wind came sweeping down the laurel-walk, and trembled through the boughs of the chestnut: it wandered away—away—to an indefinite distance—it died," Charlotte Brontë wrote.[18] *Jane Eyre* reworks atmospheric Romanticism circa 1800 at an "indefinite distance," where it had faded into historical indiscernibility. It is a novel that reflects on how it is literally inspired by a Romantic poetics of air at a belated moment, when that aerial poetics could be actualized in the present only via its prosaic disappearance. At a thematic level, figures of air from weather to breath in *Jane Eyre* intermingle physical properties and material circumstances with mental, intersubjective, and even spiritual meanings. This thematic line of inquiry is already suggested by the homophone of the novel's title, the name of its hero and narrator. Introducing herself to her pupil at Thornfield, Jane has a James Bond moment. "What is your name?" she is asked. "Eyre—Jane Eyre," she replies. "Aire?" her young student responds. "Bah! I cannot say it" (122). In that the phonetic difference between Aire and Eyre is at best vanishing, Adèle, Jane's pupil, is quite correct here: there is no distinction between air and Eyre to speak of in this novel—none sayable—although some are marked graphically. Through such moments of phonetic-graphic differentiation and through the wider pattern of semantic interconnections and disjunctions established in this text between atmosphere and narration, air and Eyre (at once the name of character, narrator, and novel) enter into the type of relationship called "synomonymy" by Justin Clemens.[19] Air and Eyre, that is, are at once homonyms and synonyms, a conjunction that, as Clemens shows, can lead to a pluralization of meaning, rather than to simple verbal self-identity.

If a homonym relates two signifieds (Sd) to a single signifier (Sg), while the synonym links two separate signifiers to a single signified, then synomonymy occurs when both of these operations take place at once (see figs. 1 and 2). The event of synomonymy, Clemens writes, "paradoxically discriminates indiscernibles in and as the binding of disparities"; it thereby "creates an interdimensional 'plane' which enables previously discriminated terms to cross and

```
  Sg              Sg   Sg
  /\              \   /
 /  \              \ /
Sd   Sd            Sd
```

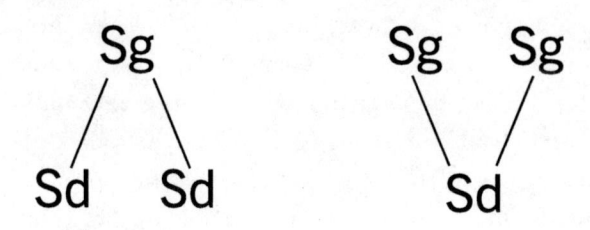

Fig. 1 Homonymy and synonymy. Courtesy of Justin Dominic Clemens.

communicate in a new distribution of the sensible." Existing close readings of *Jane Eyre* have sketched out many aspects of the semantic network that constitutes this Eyre/air (ere, ire, ear, . . .) complex of meaning. David Lodge describes the novel's structure of elemental affinities and disaffinities; Alan Bewell has written on its airs, medical miasmas, and colonial biopolitics; Justine Pizzo has traced how Brontë reworks atmo-medical discourse into a language of female self-fashioning.[20] Considered somewhat more formally, atmosphere is also a medium of literary self-reference in *Jane Eyre*. For if Jane Eyre is an artist—and the words we are reading, the very existence-in-interpretation of the text, proves that she is—then air is her chosen medium.

"Who taught you to paint wind?" Rochester demands after viewing her dreamlike, atmospheric watercolors (154). Wind here is the literal subject matter of her art; if we take the word as a double of Eyre, it is also her autonarrational subjectivity; and it is in addition, and more metaphorically, the material of her art, which as it were applies brushstrokes of wind to paper, color to air. Later, Jane writes, after Rochester's blindness, "I was then his vision. . . . He saw nature—he saw books through me; and never did I weary of gazing for his behalf, and of putting into words the effect of field, tree, town, river, cloud, sunbeam—of the landscape before us; of the weather round us—and impressing by sound on his ear what light could no longer stamp on his eye. Never did I weary of reading to him" (576–77). Although blind, Rochester is said here to be able still to see—but he sees nature and books only as translated through the sounds of Jane's voice. Earlier, Jane had painted wind, graphically representing the elusive, invisible passage of air. Now she still represents what is invisible for her blinded lover, this time through language, and the epistemological ambitions of her aero-linguistic art are evident in the fact that nature and books are treated as possible synonyms. As readers, our relationship with Jane corresponds at least in these atmospherically narrative respects to the one she has

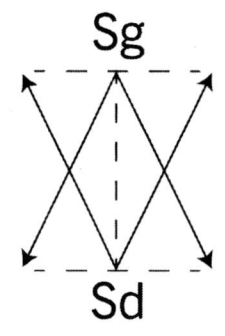

**Fig. 2** Synomonymy. Courtesy of Justin Dominic Clemens.

with the blind Rochester. For us, too, she names "river, cloud, sunbeam," calling into being the landscape before us, the weather around us, the charged moods of social encounter on almost every page of the novel. This is a transient world: rivers flow; clouds and sunbeams are impermanent and passing; the weather changes. Yet the text lays claim to a capacity to communicate the incommunicability of these passing moments. That is a paradoxical claim, in that it involves something like a determination of indeterminacy. It is also a claim asserted in and on behalf of print, which can be heard in the passage just quoted, silently, in the "press" that sounds in the middle of the "impressing" through which Jane communicates her atmospheric environment to Rochester, and to us. In *Jane Eyre*, the signs and paradoxes of atmosphere are always typographic, whatever other thematic references they might hold. Airs, in all their impermanence and elusiveness, are capable of being fixed in this novel only in the silence of the printed sign. Indeed, they could be said to have no other existence beyond that silence.

Jane discovers that she loves Rochester about halfway through the novel. Her discovery is narrated as a movement of thought that concludes with two opposing imperatives. Quoting her thoughts, she writes: "I must, then, repeat continually that we are forever sundered:—and yet, while I breathe and think, I must love him" (220). In the paragraph that closes with this sentence, Jane has been discriminating her emotions, judging her impulses of feeling, deciding what she can do, what she must do. It is a paragraph in which powerful feelings spontaneously overflow, but as the word "then," the term of inference in this sentence, reminds us, there is also a logic at work here, a practice of ratiocination or critical thinking. The precise rhetorical balancing of the sentence works to the same effect, in that it chiasmically distributes turbulent passions.

That logic divides the sentence into two equipollent halves, with the grammatical relation between the two parts indicated by the words "and yet." "Yet" is adversative, pointing to a conceptual contrast, even an opposition. In this sentence, that opposition is the deadlock inherent in the idea of an incommunicable love. "Rank and wealth sever us widely," Jane states a few sentences before (219). They are sundered irreconcilably, absolutely, unbridgeably. And yet—at the same time, nonetheless—there is also an unchallengeable identity, an unbreakable affiliation. "There is something in my brain and heart, in my blood and nerves," Jane states earlier in this paragraph, "that assimilates me mentally to him," as if her body were somehow continuous with his mind (219). This movement from body to mind—something in the blood and nerves that assimilates her mentally—is then picked up and repeated in the doublet "breathe and think," which similarly conjoins physiological metabolism and mental reflection, respiration and intellection. Jane is saying something here—to herself, but also to her reader—about breathing across the gap: about how the social chasm that necessarily and permanently divides her from Rochester is nonetheless traversed by inseverable links, a claim asserted on the grounds that breath here is a mode of thought. The felt identity of heart and mind, breath and thought, opens up a communicative dimension across that gap. This is one side of the paradoxical logic of Jane's reasoning that leads to her categorical imperative: I must love him. Against it is balanced the opposed imperative of the sentence's first clause, the logic of separation, of being sundered forever: I must not love him.

The sentence hinges at the silent points of punctuation where these two opposing imperatives meet. That punctuation is not vocalized when the sentence is read aloud. Instead, it marks the time of a pause in which something is not said, but through which something is nonetheless communicated, at least potentially. Contemporary writers on English style tend to rule out the particular typographical construction used here, the colon-dash. The purpose of the colon has now been stabilized in English written prose as indicating apposition: what follows the colon should explain or elaborate what precedes it. The dash, meanwhile, has come to signal something like a parenthesis, marking a digressive interruption to the syntactic continuity of the sentence. Colon-dashes are then to be avoided because colons indicate continuity and meaningful apposition, while dashes conversely indicate disjunction or disruption. When Theodor Adorno compared punctuation with traffic lights, the exclamation mark was red, the colon green, and the dash "halt," or amber.[21] Understood in these twentieth-century and still-current terms, the punctuation in this sentence flashes conflicting signals, saying "stop" and "go" in the same breath. It means: what follows explains or expands on what I have just said. And it also means: what follows is separate and distinct from what I just said.

These opposing functions of the colon and the dash were already present at the time Charlotte Brontë was writing. But typographical conventions were more fluid then, at least in some ways, and this particular construction was in fact quite standard in early Victorian prose style. Nonetheless, Brontë was writing during what historians of language now argue was a transition between two broad systems of punctuation: from what is sometimes called rhetorical or elocutionary punctuation to the now normative grammatical punctuation.[22] We might think of the aim of elocutionary punctuation as guiding the reader in reading the text aloud, at least imaginarily. That type of punctuation scripted how a text should be delivered, not unlike the commas sometimes found in musical scores. A colon-dash then indicated something like an extra-long pause, the pause of a colon followed by that of a dash. But the meaning of punctuation was changing, as its function of signaling grammatical and syntactic relationships became increasingly dominant. This shift in punctuation from scoring how to verbalize a text to marking abstract relations of ideas took place over a much longer period than just the nineteenth century. It was already well under way in the seventeenth, and instances of elocutionary punctuation can doubtless still be found today. And yet, much more than is the case today, Brontë was writing when both these systems of punctuation were still in play: when it was possible for a punctuation mark to mean something both rhetorically, indicating how a written sentence should be spoken, and also grammatically, to indicate its syntactic and conceptual organization.

Through the eighteenth century and into the nineteenth, these two theories and practices of the function of written punctuation coexisted relatively unproblematically. In *A Grammar of the English Tongue* (1711), John Brightland wrote, "The Use of these Points, Pauses, or Stops, is not only to give a proper Time for Breathing, but to avoid Obscurity, and Confusion of the Sense in the joining Words together in a Sentence."[23] Punctuation was a matter of both breath and sense; it registered the bodily rhythms of respiration at the same time as it governed syntactic relationships of meaning. And this collocation of both breath and sense still held into the 1840s, when Brontë was writing. The central shift—the recognition that these two systems could come into contradiction and that in both theory and practice punctuation had therefore been radically unsystematic—came with the publication of John Wilson's *A Treatise on Grammatical Punctuation* in 1844, which clearly articulated for the first time the principle that came to prevail: "the sense and the grammatical form of the construction of a passage, and not the rhetorical mode of its delivery, is the fundamental law by which the art of punctuation should be regulated."[24] The purpose of punctuation, in other words, was to mark grammatical forms to the exclusion of spoken and respiratory rhythms. Given that newly restricted purpose, the colon and the dash came to indicate conceptually irreconcilable syntactical relationships, the colon marking logical connection, the dash signaling an absence of logical connection.

*Jane Eyre* was published right at this historical threshold, in which the exclusive grammatization of punctuation had been asserted but did not yet dominate. Its punctuation, then, can be quite radically ambiguous, not least because it operates in two systems that are in the midst of being disaggregated and opposed. For centuries, punctuation had meant both breath and thought, the same sign referring to those two domains almost indistinguishably. But now thought was being extracted and isolated from breath, right at the level of these silent marks on the page. And the ambiguities of Brontë's punctuation are further compounded by its publication history. Because her text was heavily amended by her editors between manuscript and print, her punctuation practice cannot even unequivocally be said to be hers. This particular colon-dash does not in fact appear in the manuscript, and the status of the punctuation as printed in 1847 relative to that of Brontë's autograph version remains a matter of considerable editorial dispute.[25]

Punctuation articulates written language, telling the reader how the words of the sentence fit together. What it told the reader up until the 1840s was that the way breath and the spoken voice fitted words together was continuous with the way grammar and logic fitted words together. But this changed with the introduction at the level of inscription of a strong conceptual distinction between breath and thought. From 1847 or thereabouts, breath, which is a material, bodily, atmospheric dimension of language, was no longer marked (or at least not normatively or conventionally) in writing. One might almost think of this as punctuation falling silent. There is an audibility to much eighteenth-century punctuation; when we read texts from that period, breathing at all the commas, a quiet respiratory patterning becomes perceptible. But punctuation lost this somatic and mimetic function through the nineteenth century, instead coming exclusively to mark abstract relationships of logic and grammar. *Jane Eyre* stands on the cusp of this break in the system of linguistic notation, this silencing of breath in written language.

Adorno linked desynonymization, the interpretive act that generates homonyms, to the dialectic of enlightenment: "Odysseus discovers dualism when he learns that an identical word can mean different things."[26] Homonyms in this regard register the historically recurrent event of a progressive separation between the universe of sense, which is the matter out of which all words are made, and the conceptual domain of abstract and differential meaning. In *Jane Eyre*, punctuation refers to breath and thought at the same time, two linguistic dimensions that are, right at this moment, being further disarticulated. The basic logic of connection in and as division—of two worlds being both indistinguishable and inviolably separate—appears again in the two "I musts" of the colon-dash sentence, which express the conflicting demands of love and of the social conventions that forbid that love. Jane knows she loves Rochester because the coincidence of her breath and his spiritual identity tells her so: because, in this novel, passion is a form of reason, and thinking is bodily, a matter of the heart and brain, blood and nerves. But Jane also knows she cannot love Rochester, because the grammar of social relations prevents it. There is a connection across breath and thought—and a simultaneous disconnection. The historically specific sign of that paradoxical non-identity in identity is the silent colon-dash, which marks the typographic synomonymy here of thought and breath, of the conceptual realm and its nonconceptual and atmospheric exterior.

Climate change, as many have observed, appears to detach atmospheric knowledge from atmospheric sensation. Climate is global, a statistical construction of highly abstract and mediated modes of knowledge. Whatever the weather around you at any given moment, it is never climate, let alone climate change. And because our knowledge of climate change relies on what Sheila Jasanoff has described as "techniques of aggregation and deletion, calculation and comparison that exhaust the capacities of even the most meticulously recorded communal memories," scientific representations of climate change can appear difficult to reconcile with lived experience and everyday understanding.[27] It is as if we can think about climate change, but not feel it. And as if, because we cannot feel climate change, we cannot think about it too well either. The modern desynonymization of atmosphere into an object of knowledge and a communicable matter of feeling then presents a cognitive impasse or stumbling block, for climate change also appears to stage a wholesale collapse between the types of binary oppositions I have been discussing here. Because of its inescapable dualisms, the language of the present seems inadequate in this regard to the Anthropocene, understood both ontologically as a qualitatively postnatural condition and epistemologically as a grand challenge to the existing order of knowledge.

In both these registers, I have argued, the Anthropocene demetaphorizes old figures of literary textuality. But it also helps bring into focus the fact that those figures were never quite the metaphors we once thought them to be. The event of synomonymy, for Clemens, allows "previously discriminated terms to cross and communicate in a new distribution of the sensible." But synomonymy can happen backwards too, not just to terms subjected to prior acts of discrimination, but also to terms still in the act of being discriminated, on the verge of desynonymization and yet not quite across it. Atmosphere around 1800 presents one site of such a retrospective historicist synomonymy. Its transformed distribution of the sensible and the conceptual can still be registered, at an indefinite distance, in the narrative airs that punctuate *Jane Eyre*. Even the juxtaposition of a colon and a dash effects the conjunction there of atmospheric discrimination and atmospheric indiscernibility; it binds together breath and thought, passion and reason, inscriptive mark and communicative spirit, while also marking these domains as unbridgeably disparate. Rochester "saw nature—he saw books through me." Reading this book, *Jane Eyre*, we are like Rochester in that we also see only through Jane's narration, as an ear to her silent air. The typographic construction of Brontë's phrase here locates it at the historical interface between natural process and linguistic expression that I have described as atmospheric Romanticism. It is punctuated with a dash of hesitation or suspension between books and nature, which could be disambiguated into both a colon and a dash, interpreted as a connection and as a break. And yet it also erases that distinction, inasmuch as it allows both of these possibilities, the connection and the break, to be recognized as written marks of now-silenced breath. It measures our distance from the atmospheres of Romanticism:—and our distance, too, from the Anthropocene present.

## Notes

1. Raymond Williams, *Marxism and Literature* (Oxford: Oxford University Press, 1977); Gillian Russell, "'Who's Afraid for William Wordsworth?' Some Thoughts on 'Romanticism' in 2012," *Australian Humanities Review* 54 (2013): 66–80.

2. Dalia Nassar, *The Romantic Absolute: Being and Knowing in Early German Romantic Philosophy, 1795–1804* (Chicago: University of Chicago Press, 2013).

3. Robert B. Pippin, *After the Beautiful: Hegel and the Philosophy of Pictorial Modernism* (Chicago: University of Chicago Press, 2013).

4. Two influential arguments to this effect are Walter Benjamin, "The Concept of Criticism in German Romanticism," in his *Selected Writings*, vol. 1, 1913–1926, ed. Marcus Bullock and Michael W. Jennings (Cambridge: Harvard University Press, 1996), 116–200; and Philippe Lacoue-Labarthe and Jean-Luc Nancy, *The Literary Absolute: The Theory of*

*Literature in German Romanticism* (Albany: State University of New York Press, 1988).

5. G. W. F. Hegel, *Aesthetics: Lectures on Fine Art*, trans. T. M. Knox (Oxford: Clarendon, 1975), 1:54 (hereafter cited parenthetically in text).

6. Robert Kaufman, "Aura, Still*," *October* 99 (2002): 45–80, 51.

7. Friedrich Schlegel, *Philosophical Fragments* (Minneapolis: University of Minnesota Press, 1991), 39.

8. Immanuel Kant, *Critique of Judgment* (Oxford: Oxford University Press, 2007), 135.

9. Sianne Ngai, "Our Aesthetic Categories," *PMLA* 125 (2010): 948–58, 952.

10. Sianne Ngai, *Our Aesthetic Categories: Zany, Cute, Interesting* (Cambridge: Harvard University Press, 2012), 29.

11. Yuriko Saito, "The Aesthetics of Unscenic Nature," *Journal of Aesthetics and Art Criticism* 56.2 (1998): 101–11; John Barrell, *The Dark Side of the Landscape: The Rural Poor in English Painting*, 1730–1840 (Cambridge: Cambridge University Press, 1980); Ann Bermingham, *Landscape and Ideology: The English Rustic Tradition*, 1740–1860 (Berkeley: University of California Press, 1989).

12. Noel Castree, "The Anthropocene and the Environmental Humanities: Extending the Conversation," *Environmental Humanities* 5 (2014): 233–60.

13. Hans Robert Jauß, *Literaturgeschichte als Provokation* (Frankfurt am Main: Suhrkamp, 1970).

14. Ron Broglio, *Technologies of the Picturesque: British Art, Poetry, and Instruments, 1750–1830* (Lewisburg: Bucknell University Press, 2008); Arnaud Maillet, *The Claude Glass: Use and Meaning of the Black Mirror in Western Art* (New York: Zone, 2004).

15. Raymond Williams, *The English Novel from Dickens to Lawrence* (Frogmore, England: Paladin, 1974), 51.

16. Craig Martin, "The Invention of Atmosphere," *Studies in History and Philosophy of Science A* 52 (2015): 44–54.

17. Johann Gottfried Herder, *Outlines of a Philosophy of the History of Man* (London: Joseph Johnson, 1802), 418.

18. Charlotte Brontë, *Jane Eyre*, ed. Jane Jack and Margaret Smith (Oxford: Clarendon, 1969), 319 (hereafter cited parenthetically in text).

19. Justin Clemens, "Pot Tent Shell Litter Rat Shore," *Southerly*, August 28, 2015, http://southerlyjournal.com.au/2015/08/28/pot-tent-shell-litter-rat-shore.

20. David Lodge, "Fire and Eyre: Charlotte Brontë's War of Earthly Elements," in his *The Language of Fiction: Essays in Criticism and the Analysis of the English Novel* (London: Routledge and Kegan Paul, 1966), 114–43; Alan Bewell, "*Jane Eyre* and Victorian Medical Geography," *English Literary History* 63 (1996): 773–808; Justine Pizzo, "Atmospheric Exceptionalism in *Jane Eyre*: Charlotte Brontë's Weather Wisdom," *PMLA* 131 (2016): 84–100.

21. Theodor W. Adorno, "Punctuation Marks," trans. Shierry Weber Nicholsen, *Antioch Review* 48.3 (1990): 300–305, 300.

22. Cecelia Watson, "Points of Contention: Rethinking the Past, Present, and Future of Punctuation," *Critical Inquiry* 38.3 (2012): 649–72; Park Honan, "Eighteenth and Nineteenth Century English Punctuation Theory," *English Studies* 41 (1960): 92–102; Anne Toner, "Seeing Punctuation," *Visible Language* 45.1–2 (2011): 5–19; Malcolm Beckwith Parkes, *Pause and Effect: An Introduction to the History of Punctuation in the West* (Berkeley: University of California Press, 1993).

23. John Brightland, *A Grammar of the English Tongue* (1711; repr., London: 1728), 126.

24. John Wilson, *A Treatise on Grammatical Punctuation* (Manchester, 1844), 17.

25. Daniel P. Deneau and M. Thomas Inge, "The Copyediting of Literary Manuscripts," *PMLA* 117 (2002): 128–30.

26. Max Horkheimer and Theodor W. Adorno, *Dialektik der Aufklärung: Philosophische Fragmente* (Frankfurt am Main: Suhrkamp, 1984), 79 (my translation).

27. Sheila Jasanoff, "A New Climate for Society," *Theory, Culture, and Society* 27 (2010): 233–53, 249.

# 5.

# Romancing the Trace
## Edward Hitchcock's Speculative Ichnology

*Dana Luciano*

One Sunday in March 1835, prehistory put its foot down right in the middle of a quiet town in the Connecticut River valley. Its reverberations remain with us to this day. On that Sunday, W. W. Draper, a resident of Greenfield, Massachusetts, noticed some curious markings in the paving stones over which he was walking on his way home from church. He showed them to the owner of the house where they were laid, and the owner called in a local scholar, Dr. James Deane, to confirm their antiquity. Deane, who believed the marks to be fossilized animal tracks, sent a copy of them to Edward Hitchcock, then a professor of chemistry and natural history at Amherst College and the director of the Massachusetts Geological Survey. Hitchcock was enthralled by the discovery; he went to investigate and published a groundbreaking paper on the prints just a few months later, in January 1836, in the *American Journal of Science and Arts*, declaring them to be the tracks of a long-extinct species of gigantic wading birds, hitherto unknown.[1] In the wake of Hitchcock's paper, the tracks were analyzed by many of the era's most eminent geologists, including Hitchcock's mentor Benjamin Silliman, Charles Lyell, Richard Owen, and Louis Agassiz. The tracks also made their way into writings by Henry Wadsworth Longfellow, Herman Melville, Henry David Thoreau, James Russell Lowell, Oliver Wendell Holmes, Thomas Wentworth Higginson, and others.[2]

The popularity of these fossil traces, both within and outside the new field of paleontological inquiry that Hitchcock initially named "ichnolithology" (though William Buckland's term, "ichnology," was the one that stuck), demonstrates that they spoke vividly to many about the questions and possibilities emerging from the new science that made them legible.[3] Such footprints had, of course, been noted before, but what they said to Hitchcock and his contemporaries resulted from the synthesis of two still-new revolutions in geological thought, the late eighteenth-century developments that historians of the science see as its modern foundations: the recognition of the Earth's high antiquity (what we now know as "deep time") and the confirmation of species extinction, attributed to James Hutton and Georges Cuvier, respectively. In the wake of these developments, fossils came to embody the mind-numbing vastness of geological time, edged by the black border of death. Hence the fossil tracks under the feet of Greenfield's residents, even as they fascinated so many, also threatened to bring the specter of oblivion a little too close for comfort.

Evidently, these discoveries, now more than two centuries old, have not stopped happening to us. Now that humans may be on the verge of officially identifying our own geological agency, we have seen a marked critical return to concepts drawn from early geological thought, from the period following the break with biblical chronology, as the scene of conceptual innovations that we might be able to make new, or renewed, use of in our present predicament. Geologists in this period rapidly invented a radically new understanding of the Earth, one that remapped our relationship to it, and this is precisely the sort of reinvention that the Anthropocene seemingly demands. Mark McGurl tracks recent developments under the somewhat sardonic sign of the "new cultural geology": a "range of theoretical and other initiatives that position culture in a time-frame large enough to crack open the carapace of human self-concern, exposing it to the idea, and maybe even the fact, of its external ontological preconditions, its ground."[4] Geology per se is not required for this operation, though it makes a convenient critical tool in this respect.[5] Yet spatiotemporal vastness is not the only aspect of geology, new cultural or otherwise, to which this emergent body of thought appeals. More germane to my purpose in this chapter is the protogeological understanding of inorganic (and, in the case of the fossil, postorganic) matter as animate and self-organizing that we see in new materialism.[6] There, the geological tends to operate not simply as a gateway to a vastness that explodes the human but, perhaps more usefully, as a way for humans to find themselves otherwise—to find themselves *mattering* in relation to the histories of material agency that geology explores.[7]

As Noah Heringman has demonstrated, although early geologists insisted, contrary to their biblically influenced predecessors, on empirical and rational study, they were not immune to speculative flights of thought, to aesthetic and sensory enchantment by geological matter, and to philosophical musings drawn from the rocks.[8] The "aesthetic," understood as the "sensuous consideration of what is indeterminable in things,"[9] opens geology to speculation by means of what Heringman terms "wonder": a mode of invested inquiry that exceeds and redirects the empirical. Wonder, as distinct from awe, does not annihilate the subject so much as redirect it. I focus in this chapter on the Romantic wonder that surfaces throughout Hitchcock's writing on the Connecticut Valley fossil tracks, leading to what we might identify as an "ichnopoiesis" articulating itself alongside and through his extensive empirical study of the tracks.[10]

Hitchcock did not invent the field of ichnology, the study of fossilized traces (footprints, claw and teeth marks, and so on), nor did he discover the strange-looking marks on which he made his reputation, but he was the first scholar to geologize them in print. A commission sent by the American Scientific Association in 1841 vindicated his conclusions; that same year, Hitchcock's second published geological survey of Massachusetts, which included a lengthy section on the prints, was so popular that Thoreau, commissioned by Ralph Waldo Emerson to write on the question, could not obtain a copy because the books had sold out by the time he went looking.[11] The transdisciplinary scope of Hitchcock's writing on the tracks permits a departure from the conventional "science and literature" model, wherein the prosaic observations of the working scientist provide the raw material for the more interesting or aesthetically satisfying reflections of the philosopher or poet.[12] Since Hitchcock wrote on the fossil footprints in all of these modes (as a scientist, poet, and theologian), he resists easy assimilation to critical models that assume a bifurcation between the arts and the sciences, a division that was, in his day, still emergent. As both a working scientist and a practicing clergyman, Hitchcock also troubles the narrative of secularization, the breaking free of biblical chronology and overcoming of Christian opposition to science that, as I have observed elsewhere, has effectively come to be written *as* the history of modern geology[13]—which we might question in tandem with the disciplinary division mentioned above, insofar as academic secularization went hand in hand with the kind of professionalization that sorts science and literature into separate categories. Hitchcock's ichnopoiesis, in this light, raises the possibility of an undisenchanted geology, one whose energies, I suggest, remain necessary in our own time.

Although Hitchcock's fragmentary musings on the moral implications of fossil footprints do speak suggestively to certain tendencies in contemporary new materialist thought, my intent here is to set up a speculative cross-temporal conversation, not to establish a critical genealogy or identity; Hitchcock is not a forerunner of new materialism, but rather an intriguing interlocutor. Nor do I seek to uncover anything like a unified ethical theory or plan of action in Hitchcock's work, because there isn't one. Indeed, the messiness of his musings is part of what draws me in this context—because there, too, I find points of contact with contemporary new materialist thought, which has been less successful in generating a politics in this sense. For the sake of organization, though, I have divided Hitchcock's speculations into two parts: one considers the implications of ichnology in something like an organicist frame, under the sign of the tree of life and in the form of poetry; and the second examines ichnology in an inorganic frame, under the sign of the calculating engine and in the mode of theology. But first, a bit of background on the nineteenth-century interest in fossil traces.

## Ichnomania

The sort of markings that fascinated Hitchcock and his contemporaries had been interpreted by earlier observers. One Lenni Lenape account held that the marks were made by an ancient race of monsters that preyed on all other living creatures until they were destroyed by lightning. The Haudenosaunee used them to aid the growth of crops. Residents of South Hadley, Massachusetts, found a set of prints in 1802 and identified them as the tracks of Noah's raven; W. W. Draper believed the Greenfield markings to be three-thousand-year-old turkey tracks. But the rapidity with which the marks went geological in the 1830s says a great deal about the appeal of the science to Anglo-Americans in this period. The practical uses of geology were embraced by the young republic; beginning with Lewis and Clark, geological surveys were part of all major Western expeditions, and states also appointed their own surveyors, starting with the Carolinas in 1823–24 and Massachusetts in 1830 (the latter under Hitchcock's leadership). Once theories of the Earth's antiquity began to gain wide circulation (fueled especially by the publication of Charles Lyell's accessibly written *Principles of Geology* in 1830–33), the more speculative aspects of the science began to take root in the popular imagination. By 1834, geology could be called the "fashionable science" in the pages of the *Knickerbocker; or, New-York Monthly Magazine*.[14] Geology's revelation of what Walt Whitman would call "the infinite go-before of the present" was touted as a source of fascination greater than anything humans had managed to produce.[15] As the *North American Review* put it in 1836: "The world has a history written on its strata: a history so interesting, that the most splendid fictions of the human imagination sink into insignificance when compared with it."[16]

This oft-noted insignificance did not, however, amount to an erasure of human history in the face of the new science's deeper temporal accounts; rather, the fascination with geology was to a large extent the project of locating the relationship between human and planetary history through the accounts read in the rocks. That history could offer the stability of self-confirmation through the slow, stately progress that Lyell and other uniformitarian thinkers insisted was the general shape of Earth's history, and it was used to undergird some of the Western world's most familiar self-locating accounts, including settler colonialism and white supremacy.[17] Yet geology's gaps and differences, its reflection of the deep past as "the theatre of reiterated change" and "neverending fluctuations," as Lyell also observed,[18] meant that it could press forward visions of life lived otherwise.

We can begin to understand the distinctive appeal of ichnology—literally, the science of traces—in this light. Gayatri Spivak explicates the Derridean sense of "trace" as the "mark of the absence of a presence, an always already absent present, of the lack at the origin which is the condition of thought and experience."[19] Yet in the case of fossil tracks (as Spivak observes, the French word for "trace" conveys the sense of "track" or "footprint" more strongly than the English does [xv]), we might more accurately refer to the *presence of an absence*: the mark of the here-no-longer that nevertheless remains. This reversal recalls Steven Shaviro's Deleuzian reading of the positive inflection given the Derridean trace in his late work "Specters of Marx," where the trace reappears as a "radical non-negativity, a kind of residual, quasi-material insistence, that disrupts and ruins every movement of negation or negativity."[20] As a specter, the trace haunts the present not with an absence that ruptures presence, but with a presence that negates negation.

All fossils, in this sense, are traces: lithic ghosts incapable of disappearing; material echoes of past life; forms that, by refusing to vanish into the abyss of time, prevent time from becoming merely abyssal. This is true pragmatically as well as philosophically: stratigraphy, the method of mapping the geological past developed in the early nineteenth century by William Smith, relies to a large extent on the correlation of index fossils and of mineral materials for determining the relative age of strata. Even here, though, the traces seem to embody the paradox of lithic persistence, the possibility of having one's death and surviving it too. On this view, the term "trace fossil," as it designates the subject of ichnology, might be regarded as redundant. Yet the repetition serves to emphasize an intensified form of persistence past extinction indexed in the prehistoric footprint. The fossil trace displaces the morbidity embodied in other fossils—that is, while the death of the creature that made it can be assumed, it is not present in the form of an ossified body. While ordinary fossils offer a record of existence, of being or having been, fossilized traces, and footprints in particular, appear as a record of action, of doing or having done. They seem to open directly onto a dynamic (rather than merely taxonomic) vision of past worlds. Ichnology, as the present-day trace fossil specialist Anthony J. Martin contends, is "[p]erhaps more than any other part of paleontology . . . about that exciting intersection between science and flights of fancy."[21] Because it demands the speculative conjuring up of a being from the impression of its presence—rather than working to piece together its remnants—ichnology can offer a fuller, fleshier picture of a creature interacting with a physical world that was likewise in flux. As it tracks the traces of creatures in motion, then, ichnology activates the geological past as a resource for speculation and sensation, not as a vehicle for annihilation.

**Fossil Poetry**

Hitchcock was not slow to turn the Connecticut Valley prints toward speculative ends. In December of the year that his research on the fossil footprints first appeared, the *Knickerbocker; or, New-York Monthly Magazine* printed a poem with the title "Ornithichnites Giganteus, *Redivivus*." The poem was published under the pseudonym "Poetaster," a term designating an inferior dabbler in the poetic arts, which had also by Hitchcock's time come to mean one who wrote on mundane subjects. Hitchcock wrote this poem, though his authorship was never revealed, and both the journal's editors and the prefatorial correspondence from Poetaster, which established Hitchcock's *American Journal of Science and Arts* article as the inspiration for the poem, invoked him only in the third person.[22] The *Knickerbocker*'s editors nodded to the article's conclusions without embracing them. Hitchcock-as-Poetaster, though, actively promoted them, affirming, "there was at least probability enough in the theory advanced . . . [by Hitchcock] to make it lawful to use it in verse."[23] To a twenty-first-century reader, the selection of a "lawful" subject for poetry based on scientific probability might seem beside the point. It makes sense, though, in the poetic tradition upon which Hitchcock draws, that of scientific didactic poetry, where the poet serves not only as the conveyor of versified information to the reader but as an authoritative editor of such information, selecting among competing theories to present the most accurate facts.[24]

The *Knickerbocker* poem presents three speakers in succession. The first, a solitary geologist examining a set of fossil tracks, muses on the enormous changes the Earth has seen since they were made. He enjoins the creature that left them to return, lamenting that modernity has dissipated the magicians who might make that happen. In response, a sorceress appears. Declaring science to pose no obstacle to her power, she causes a huge bird to burst out of the rock in a torrent of water. The bird, identifying itself as the source of the prints, scorns the sorceress's equally mocking description of "Man," a creature scarcely six feet high, as "creation's lord." Claiming that title for himself, the bird taxonomizes the amazing creatures of his epoch down from his own ruling position. Once finished, he refuses to remain in the geologist's cold and degenerate world:

> Sure 'tis a place for punishment design'd;
> And not the beauteous, happy spot I lov'd;
> These creatures here seem discontented, sad;
> They hate each other and they hate the world.
> O who would live in such a dismal spot?

The bird sinks back into the rock, and the "vex'd geologist" is left grasping after its form. He laments the loss to science and curses the arts employed by the sorceress,

> Forgetting that the lesson taught his pride
> Was better than new knowledge of lost worlds.

The close of the poem reveals its didactic content to be hybrid: part scientific and part moral. Yet while the geological and paleontological lessons offered in the speech of the geologist and the bird are clear (if now outdated), the moral teachings overlaid on this scientific substrate, the lesson missed by the geologist, is less so. If we take the bird's speech at face value, the poem reads as a rebuke to the geologist's world, which the bird describes in quasi-apocalyptic terms, calling it "well nigh worn out" and predicting that its occupants are "soon / in nature's icy grave to sink for aye." The poet, in a footnote, acknowledges the bird's view as a potentially rational conjecture in light of the evident cooling of the globe. We might, then, read the poem as a remnant of diluvialist views of the pristine prehistoric past in contrast to the broken world in which we now live; the kind of history presented in biblically based geological writing, such as Thomas Burnet's *Telluris Theoria Sacra* (1681).[25] Earth history, here, works primarily as a shattering of human hubris, and the sandstone footprints are a planetary manifestation of the Puritan death's head, a lithic memento mori.

Yet the target of the "lesson" may not, after all, be human hubris, but scientific haste. Hitchcock's geology was more modern than Burnet's by far; he embraced both the post-Huttonian geological time scale and Cuvier's proofs of species extinction, though he refused theories of evolution, along with Cuvier and his own contemporary Agassiz. The geological past appeared, in Hitchcock's view, as a series of perfectly integrated ecosystems. In his *Elementary Geology* textbook, first published in 1840, he depicted this history in a paleontological chart based on the then-common trope of the tree of life (fig. 3). That the diagram looks more like a bush than a tree reflects Hitchcock's conviction that all major plant and animal species had been present since the beginning of life, though they had become more sophisticated and more diverse over time—hence the spreading of the branches. His positing of birds as the makers of the Connecticut Valley tracks was crucial to this picture. As noted in another footnote to the *Knickerbocker* poem, the lack of any bird fossils far down in the rock record had led many geologists to infer that they did not exist then. The discovery of their footmarks, as Hitchcock had argued in the *American Journal of Science and Arts* article, might therefore "prove . . . an instructive lesson to the geologist," leading him to "enquire, whether he has not been too hasty in inferring the non-existence of the more perfect animals or plants, in the earlier times of our globe" ("Ornithichnology," 340). Moral didacticism, in this reading of the poem, would be folded back into scientific didacticism to correct an overreliance on empiricism, a lack of principled speculation.

Both these readings, though, overlook the fleeting but crucial appearance of the sorceress in the poem. Hers is a strange interlude, set apart metrically from the matching styles of the geologist and the giant bird; they speak in the pentameters favored in Anglophone scientific didactic poetry, while her brief speech, in rhyming tetrameter, echoes Shakespearean invocations of the

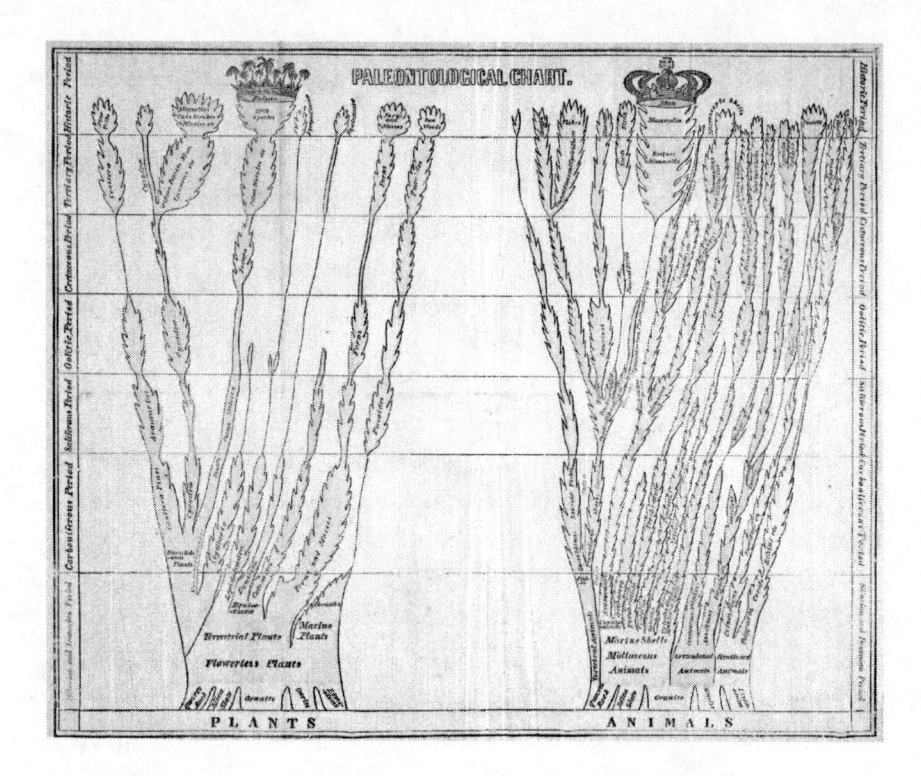

**Fig. 3** "Paleontological Chart." From Edward Hitchcock, *Elementary Geology: A New Edition, Revised, Enlarged, and Adapted to the Present Advanced State of the Science, with an Introductory Notice by John Pye Smith*, 30th ed. (New York: Ivison and Phinney, 1857). Photo: Princeton University Library.

supernatural. Yet even as her arts accomplish what the empirically bound geologist cannot, she herself has been conjured inadvertently by him, since his wish brings her into the poem—and crucially, that wish also contains an allusion to the story of the witch of Endor. (The story, from the first book of Samuel, runs thusly: King Saul, about to face the Philistine army, enjoins the witch to summon the spirit of the prophet Samuel. The spirit appears, very cranky about having been disturbed, chastises Saul for disobedience to the Lord, and prophesies his downfall, which takes place as promised the following day.) Much commentary on this incident holds that its point is *not* Samuel's prophecy (indeed, his appearance is said to be a trick, since the dead, in Judeo-Christian theology, do not usually return, with one significant exception) but Saul's hypocrisy, since he had banished sorcerers from the realm. The poem may likewise condemn the geologist for hypocrisy, since he claims at the outset that such enchantments are banished from the modern "age of light." Summoning the sorceress is thus both hypocrisy and scientific impiety.

A strictly biblical reading of the poem is complicated, though, by the sorceress's Romanticism: her metric alterity, her scorn (cut in the published version) at the scientific arrogance of naming what remains inaccessible, and most important, the visual and poetic revisions of the witch by Romantic writers and painters (Henry Fuseli, Blake, Byron), which the poem might have brought to mind for nineteenth-century readers. For Romantics, the witch defiantly maintains a place in a present that would banish her but cannot, precisely because she occupies a site less supernatural than psychic: the site of interiority, the turbulent and half-occluded domain of the emotions that surround our desire for and fear of the past. In this context, then, the sorceress is linked less strictly to heresy than to the potentially heretical energy that drives scientific inquiry—the wondering intensity of the geologist's deep desire to know the geological past. Wonder operates as a vitalizing force enabling the imagination to access prehistory; it is necessary to drive the labor of discovery. At the same time, it needs to be carefully controlled. As Heringman reminds us, wonder was repeatedly disavowed by scientific writers, since there is no telling what might quicken through the rocks. What we can term, with the sorceress, the "enchantment" of prehistory points toward embodied, and potentially feminized, knowledges that surpass taxonomy, although their romance with the dispossession of the modern would remain carefully regulated.

## Distributive Geology

Though one "moral" reading of the fossil footprints views them as a geological memento mori—remember, you must die—another assessment, which Hitchcock began developing in the 1840s, moved in a different direction: also remember, you may not. This caveat is not quite as triumphalist as it might at first sound; the capacity to survive across eons is not only a testament to the success of humans but also a potential testimony to their perpetual fallibility. The shift in emphasis is the result of Hitchcock's turn from the tree of life, the organic metaphor he earlier employed as a means of grappling with the past presented in the footprints, and toward the mechanistic model indexed in Charles Babbage's *Ninth Bridgewater Treatise* (1837–38).[26]

Babbage, a controversial Cambridge mathematician and inventor now recognized as an early pioneer of the computer, had composed an unauthorized addition to the eight official *Treatises on the Power, Wisdom and Goodness of God* commissioned by the Earl of Bridgewater, which explored the contributions of different sciences to natural theology. Annoyed because his Cambridge colleague and rival William Whewell had remarked in his treatise that there had been no positive moral contribution made by mathematics, Babbage sought to illustrate the possibility of using math to discern the unity and beauty of the divine plan, in part through tracing the cosmic implications of the principle of action and reaction. Matter, he asserted, was permanently affected by motion, which effectively never ceases: once sent into the air, the energy that carries a given sound remains even after the sound itself has ceased to be heard. The principle applies to liquids and solids as well: "The solid substance of the globe itself, whether we regard the minutest movement of the soft clay which receives its impression from the foot of animals, or the concussion arising from the fall of mountains rent by earthquakes, equally communicates and retains, through all its countless atoms, their apportioned shares of the motions so impressed" (Babbage, *Ninth Bridgewater Treatise*, 115).

The deep memory of the Earth, Babbage noted, is too subtle to be perceived by ordinary human senses but might be tracked mathematically. Were there a being who enjoyed perfect command of mathematics, he observed, it would be able to follow and potentially to predict the course of all motion—though even this being, "however far exalted above our race," would come nowhere near the intelligence of the deity (*Ninth Bridgewater Treatise*, 110). Though the legibility of the past and the predictability of the future in this model have led commentators such as Alan Liu to view Babbage's computational theology as a deterministic one, with the fate of the universe effectively scripted in advance, it is actually more flexible, more open to chaos and contingency. As Tina Young Choi argues, "Babbage's account focuses on change, on the shifting paths, extinctions, and redirections that result from 'altered physical circumstances' . . . all of which a superior providential plan, like a superior computational one, can accommodate."[27] Divinity, that is, signals not a refusal of chance and flux but a supremely advanced ability to cope with it.

Hitchcock first linked Babbage's speculation on matter's memory to the fossil footprints in an 1844 report in the *American Journal of Science and Arts* on the progress of ichnology as a field. The report punctuated a lengthy review of the data gathered thus far with a nod toward some of the "valuable moral[s]" that the field provided. First among these was the proposal that ichnology "shows us that the most trivial movement of ours may make an impression on the globe that shall be brought out ten thousand years hence with unimpaired freshness—that shall in fact be immortal" ("Report on Ichnolithology," 320). Hitchcock went on to posit that the fossil tracks are "almost a realization of the ingenious thought of Prof. Babbage . . . that 'the air is one vast library, on whose pages is written all that man has ever said, or woman whispered; while the waters and the more solid materials of the globe bear equally enduring testimony of the acts we have committed'" (321). The promise of immortality, the time-defying evocation of freshness: these seem to establish ichnology as a site for the transcendence of time, as perhaps the ultimate location of human mastery. Babbage's thinking, accordingly, is positioned as a more or less straightforward guarantee of the permanence of human action—so much so that Hitchcock felt the need to balance this affirmation by citing a "lesson of an opposite character" drawn from the official *Bridgewater Treatise* composed by geologist William Buckland ("Report on Ichnolithology," 321). Buckland, in that text, points to the triviality of the kind of motion that is the subject of ichnology, conspicuously contrasting the geologizing of unexceptional everyday actions—like walking in mud—to the dramatic events that humans think will have historical significance, like going to war.

Though Hitchcock set up Buckland's reflections as a corrective to the potential excesses of hubris opened by the suggestion of human immortality, Babbage's system contained its own built-in check: the kind of impression one makes cannot be known or tracked in advance, nor can it be controlled or erased. Babbage suggested toward the end of his treatise that the fact that the Earth will record *all* actions, good, evil, and neutral, ought to give us pause about the impact of our deeds. This dimension of his thinking became implicit in Hitchcock's return to the Babbage principle in his 1858 volume, *Ichnology of New England*. Hitchcock's concluding reflections, which he identifies as "curious speculations," expand the claims from the 1844 report to underscore the operation of the will alongside the taking of action: "the slightest action of ours, *even the most unnoticed decisions of our wills*, may make an impression on the globe, which will endure, and may be read, as long as the earth exists."[28] Babbage was not directly quoted, though he was referenced in a footnote. Instead, Hitchcock walked through the making of fossil footprints in order to illustrate his claim. Pointing to some of the illustrations of fossil tracks in the volume, he described in great detail how they reflect not just the presence of prehistoric birds and insects, but also the specific muscular actions they took to make particular impressions, actions legible to the trained geological eye across vast spans of time:

> To illustrate this thought, let me recur to Plate XLV., fig. 1, where Tridentipes gracilior is seen to have marked out a portion of an ellipse by its track; or to Plate XXX., fig. 3, where an insect, or crustacean, is seen to have marked out a circle by its track. Now this change in the animal's course must have required the use, and the increased or diminished action, of certain muscles in its legs. We have, then, in these curved tracks, certain evidence of the peculiar action of the Adductor, or Gastrocnemian muscle, or all of them, together, in the leg of a small bird; nay, of a small insect; perhaps ten, perhaps fifty, perhaps a hundred thousand years ago! Still further, that muscular movement implies a previous action of the animal's will, and that implies, as we now know, an electric current inward along the sensor nerve, and outward along the motor nerve. With the register before us of the decision of an insect's will, made fifty thousand years ago, and the corresponding movement in the muscles of its legs, who will dare to say that any action of ours, or any operation of the human mind, will certainly be so lost that it may not reappear in all its freshness ten thousand ages hence! (173–74)

Hitchcock's extension of the durable "moral" implications of fossil impressions from the afterlife to the inner life reflects his developing conviction that thought counts as a kind of action, and hence can also have lasting material effects. For Hitchcock the theologian, this theory held significant spiritual implications, such as the question of whether negative thoughts themselves could impact others negatively. For Hitchcock the geologist, though, the detailed description of the connections among thought or will (animal or human), the "electrical" structure of the nervous system, and the muscles of the body as they made contact with the then-plastic sandstone, alongside his previous account of the subsequent climatological and geological forces—the perimineralization, sedimentation, and erosion that led to the chance event of an impression's stony preservation—sketches a version of distributive agency across deep time. Even as he distinguished between human or animal vitality or mental activity, the physical body, and the material world, his description underscored the necessary connections among thought, energy, flesh, mud, minerals, sediment, wind, and water, emphasizing this preservation as a collaborative or compositional process, as well as a matter of chance or speculation.

The materialization of such traces of action, unlike the certainty of Babbage's system, is only a possibility, dependent on a number of factors beyond one's own agency. Moreover, the preservation of such traces is as much outside one's control as the erasure thereof—a possibility that for Hitchcock, as for Babbage, provided a kind of moral oversight on action, since anyone's false moves might likewise be preserved and read until the end of the Earth.

The possible preservation of human action across time ("immortality" is notably absent in the revision of this argument contained in *Ichnology of New England*) also necessarily emphasized the unforeseeable moral, geological, and environmental consequences of any move we might make—even, potentially, of any given mind-set. In this sense, Hitchcock's "curious speculations" anticipate similar reflections on the environmental implications of distributive agency as it has been taken up by new materialist ecocritics, who emphasize "the world's phenomena [as] segments of a conversation between human and manifold nonhuman beings," as Serenella Iovino and Serpil Oppermann put it.[29] This observation has both analytic and ethical implications, based on the recognition that we cannot entirely know the effects of our own actions. To affirm the potential durability of human action is also paradoxically to recognize a certain limit to human agency, or at least the necessity of conscious collaboration with other material entities as that action is thrown into a deep future that humans can neither predict nor control.[30] But the diminishment of human agency amounts for new materialists—as for Hitchcock—to an intensification of human responsibility: the human inability to fully predict or control the behavior of the nonhuman world ought to generate deeper ethical reflections on the potential implications of any given action.

A kind of thrown-ness is also happening to Hitchcock's writing itself, at least in this reading. Neither Hitchcock's geological nor his moral reflections coincide precisely with twenty-first-century thought about material agency, vibrancy, and the necessity of wonder, yet they may end up moving in similar directions. They also reflect, or perhaps predict, what is taking shape in present-day Earth system science as the "stratigraphy of the future," a term coined by Elizabeth Kolbert to describe the work of Jan Zalasiewicz, the scientist who heads up the International Commission on Stratigraphy's Anthropocene Working Group: a speculative attempt to read the rock record ten, or a hundred thousand, years hence to see how much of an impact humans have made. Like Hitchcock's almost offhand mention in the 1841 *Geological Survey of Massachusetts* of the human capacity to cause climate change, the large-scale environmental implications of nineteenth-century geological and ichnological thinking have begun to take on the kind of life of their own that Hitchcock suggested any human action or thought might have: however unintentionally, they have become clearer and more pronounced over time, as latter-day readers begin to make links between fossil footprints and carbon ones.

### Coda: Glowing Language and Material Consequences

As I noted near the outset of this chapter, Hitchcock's speculative engagement with the fossil tracks does not resolve into anything like a clear cosmology, much less a plan of political action. Nor, much as we might wish for one, do contemporary new materialisms. Rather, I think, what they offer is another way of engaging the world, of reorganizing the sensible through the elaboration of a newly speculative sensorium, one that has taken on a particular urgency under the sign of the Anthropocene. Still, the relationship between new materialism's affective and historical politics, between the material and the social in the precarious context of environmental crisis, remains a slippery and charged one. One final meditation from Hitchcock might help us think about the stakes of that slipperiness.

The "moral lesson" of Babbage's thinking for Hitchcock, left implicit in the scientific papers, was brought out most emphatically in *The Religion of Geology and Its Related Sciences*, which quoted in full a paragraph added to the second edition of Babbage's treatise. The passage charges, in effect, that the planet itself will remember the inhumanity of the slave trade. Hitchcock included it in a chapter titled "The Telegraphic System of the Universe," where he again cited Babbage's illustration of "the principle of reaction," declaring, "Not a footprint of man or beast is marked upon [Earth's] surface, that does not permanently change the whole globe."[31] Though we are unable to see most of these, he insisted, "in a higher sphere there may be inlets of perception acute enough to trace [them]" (Hitchcock, *Religion of Geology*, 413). He turned back to Babbage's text to drive this point home:

> In view of these facts, we cannot regard the glowing language of Babbage an exaggeration, when he says, "The soul of the [N]egro, whose fettered body, surviving the living charnel-house of his infected prison, was thrown into the sea to lighten the ship, that his Christian master might escape the limited justice at length assigned by civilized man to crimes whose profit had long gilded their atrocity, will need, at the last great day of human accounts, no living witness of his earthly agony: when man and all his race shall have disappeared from the face of our planet, ask every particle of air still floating

over the unpeopled earth, and it will record the cruel mandate of the tyrant. Interrogate every wave which breaks unimpeded on ten thousand desolate shores, and it will give evidence of the last gurgle of the waters which closed over the head of his dying victim. Confront the murderer with every corporeal atom of his immolated slave, and in its still quivering movements he will read the prophet's denunciation of the prophet king." (413)

The subject matter of the "glowing language" that Hitchcock quoted extensively was not followed up. Indeed, *The Religion of Geology* was not centrally concerned with the Atlantic slave trade or the slave system in general.[32] (Nor, for that matter, was Babbage: the passage in question was added to the second edition of his treatise in 1838, after Babbage read of cases of slavers dumping their human cargo overboard.) The slave system, in both texts, appeared as the historical nadir of the generalized systems of evil with which the authors were concerned, and the image of a recording Earth was posited as a rebuke to and redress for a legal system that failed to arrest these practices.

In the context of something like a "religion of geology," Hitchcock's adaptation of Babbage instantiates a different framing of the matter of race than the one geology was usually employed to shore up in this period. In evolutionary thought, "race" has persisted historically as a kind of species belonging, a form of biological life marked by a set of hereditary traits, moving forward across time, and hence something subject to extinction. Indeed, the use of the term as a naturalized alibi for the displacement and extermination of Indigenous populations was common in this period.[33] Henry Wadsworth Longfellow, in "To the Driving Cloud" (1845), linked the Connecticut Valley fossil footprints to the image of a Native American chief walking westward, off the historical stage: "What, in a few short years, will remain of thy race but the footprints?"[34] Hitchcock's use of Babbage, however, engaged the tracks differently, suggesting a vision of the historicity of race not as a kind of life or biological matter, but as a material relation, manifested in a series of transactions whose vibrant affective residue is archived by the inorganic material of the Earth.[35] In this sense, Hitchcock's geologizing of the passage from the *Ninth Bridgewater Treatise* permits us to see race not as narrowly biological but as broadly ecological, determined in this instance by the violence of the slave system in its dehumanizing construction and disposal of Africans as "waste material."

In light of this reframing, I want to briefly note two points about the performative impact of this passage, ignited by a colleague's comment that it "feels right, but it isn't true." Now, by saying it "feels right," she was not being dismissive, in the way that we have been trained to think of that sensation as it appears in numerous critiques of the political inertia of nineteenth-century white sentimental writing. Rather, she was pointing to something akin to what Caleb Smith calls a "poetics of justice": these "glowing words," to use Hitchcock's language, are "capable," in Smith's, of "summoning readers who *burned* . . . to remake their common world."[36] There are intriguing points of overlap between the way Smith frames nineteenth-century martyr literature and the contemporary moral (re)activation of matter: in particular, Smith's observation that martyr literature is not a genre of self-aggrandizement, but one of self-abnegation or desubjectivization (a central goal of much inhumanist thought), and Smith's identification of such writing with the search for a counterpublic animated by religious or ethical structures of feeling rather than legal ones (a sentiment that also resounds in contemporary ecocritical and new material discourse). The desire for a more vibrant connection, not only between human and (non)human matter, but also between cultural and critical texts and their readers, is palpable in new materialist work; this desire can be felt in its broad disparagement of the linguistic turn and its supposed affective slenderness. While I disagree with those critiques of the linguistic turn that fault it for its anthropocentrism—insofar as the poststructuralist view of language was that it was essentially inhuman—there is something to be said for this desire. In an ambivalent discussion of the nonhuman turn, Mark Seltzer identifies its emergence as part of "a reaction against . . . the zombielike aftermath of deconstruction."[37] The emphasis on enchantment as an effect of close and slow reading, on the interpretation of texts and matter as, in Jane Bennett's terms, a "cultivated, patient, sensory awareness,"[38] responds to this critical condition, working against the deadening of reading by trying to make the world come alive again. In this respect, it may become a vitalizing complement to the project Smith describes—though only if, as Hitchcock's reframing of Babbage suggests, it attends to the traces of the human within nonhuman matter as well as the other way around.[39]

Also central in this respect is my colleague's observation that the passage Hitchcock cited "isn't true" in a material sense. This assertion is at odds with Hitchcock's insistence that these material impressions are detectable only from the superior sensory perspective of "a higher sphere"; in other words, from his perspective, we do not yet know, since we have not yet ascended there. Even if we remain earthbound in our quest for truth, my colleague's objection is belied by a number of cosmologies which hold that histories of social and ecological damage permeate the matter of the planet. This view has been entered into geological thought by a study that appeared in the journal *Nature* in early 2015, authored by Simon L. Lewis and Mark A. Maslin, which proposed a new start date for the Anthropocene (1610) based on a dip in carbon levels detectable in Arctic ice cores.[40] This dip, they assert, operates as a material, geological marker of the deaths of at least fifty million Indigenous residents of the Americas in the first century after European contact, an event inseparable from the subsequent ecological and geological damage that the epoch seeks to mark. Responses to their proposal have revealed the divide between the humanities and the sciences with respect to Anthropocene thought; while it sparked affirmative reactions within the humanities,[41] it ignited a firestorm of criticism in the pages of scientifically focused Anthropocene journals. Twenty-two members of the Anthropocene Working Group wrote a sharp rebuttal of the proposal, which charged not only that the scientific evidence Lewis and Maslin offered was insufficient, but also that the kind of history they proposed was not the point of the epoch: the "choice of human narrative" should not guide the selection of Anthropocene boundaries, but rather "the positioning of a stratigraphic boundary should simply be pragmatically and dispassionately chosen, by the same manner in which all earlier stratigraphic boundaries were chosen, to allow the most effective practical division."[42]

From a contemporary scientific perspective (the Anthropocene Working Group is overwhelmingly dominated by scientists), this assertion contradicts what we might call the *spirit* of the Anthropocene, at least as it has developed in the popular imagination, in two ways. First, its insistence on a "dispassionate" manner of boundary selection identical to previous ones negates both the polemical intent of the epoch and its radical difference from all previous epochs. Second, as both Hitchcock and Bennett emphasize, from the perspective opened by the "stratigraphy of the future," we *cannot yet say* what will become geological evidence. From this perspective, then, we might speculate how this and other proto–environmental crisis factors, such as the toxicity of labor exploitation and racial violence, will be legible in that stratigraphy, and we might, accordingly, begin to refine the kinds of material reading or sensing practices we use to detect these factors at present. (This may well require more attention to things like deadness and negativity than some new materialist thinkers are yet comfortable with—though one beneficial outcome of a sensorily attentive practice of historical reading, I propose, would be an undoing of the opposition between positive and negative poles of affect or power.)

Such methods may not pass muster with those who currently assert the priority of dispassionate and objective modes of analysis in both the sciences and the social sciences, nor with those who advocate the "descriptive method" in the humanities. Yet far from overcoming C. P. Snow's "two cultures" divide, a task that decolonial theorist Sylvia Wynter, who values humanistic thought more than Snow does, identifies as essential to developing an ecologically attentive and globally humanism (one that is not yet in existence), the overvaluation of empiricism as detachment ultimately upholds it.[43] In this sense, we need to remain invested in attachment and passion, even as we move to encompass the nonhuman within these investments. I am not calling for critical reenchantment, since I remain wary of what that nexus of desire may overlook; the more balanced attentions of *undisenchantment* may better serve our purposes as we move into the new "crisis ordinary" of the Anthropocene.[44]

## Notes

I am grateful to audiences at Columbia, Rutgers, Cornell, Johns Hopkins, SUNY–Albany, and the University of Toronto for their responses to this material. A short essay drawing on this archive, "Tracking Prehistory," was published in J19: The Journal of Nineteenth-Century Americanists 3.1 (2015): 173–81; thanks to John Lardas Modern and Christopher Castiglia for their editorial responses to that piece.

1. Edward Hitchcock, "Ornithichnology— Description of the Foot Marks of Birds (Ornithichnites) on New Red Sandstone in Massachusetts," *American Journal of Science and Arts* 29.2 (1836): 307–40 (hereafter cited parenthetically in text). This first article, published just months after the discovery, is still recognized today as a masterly work, even though many of its conclusions have been abandoned. See William Miller, *Trace Fossils: Concepts, Problems, Prospects* (Amsterdam: Elsevier, 2007), 49: "I find Hitchcock's inaugural 1836 monograph awe-inspiring."

2. See D. R. Dean, "Hitchcock's Dinosaur Tracks," *American Quarterly* 21.3 (1969): 639–44, for a review of some of these works.

3. Edward Hitchcock, "Report on Ichnolithology; or, Fossil Footmarks," *American Journal of Science and Arts* 47.2 (1844): 292–325 (hereafter cited parenthetically in text).

4. Mark McGurl, "The New Cultural Geology," *Twentieth-Century Literature* 57.3–4 (2011): 380–90, 380.

5. McGurl's primary example is Quentin Meillassoux's discussion of the arche-fossil in *After Finitude: An Essay on the Necessity of Contingency*, trans. Ray Brassier (New York: Continuum, 2008).

6. In this sense, I am reading against the tendency to group new materialism with object-oriented ontology, speculative realism, and other emergent areas of thought in critiques of the geological/ontological/nonhuman turn, not only by McGurl but also by Christopher Nealon, "Infinity for Marxists," *Mediations* 28.2 (2015): 47–63; and Jord/ana Rosenberg, "The Molecularization of Sexuality: On Some Primitivisms of the Present," *Theory and Event* 17.2 (2014), https://muse.jhu.edu/article/546470, with which I am otherwise largely in sympathy. I do this because I find the often explicitly feminist emphasis on the embodied and sensory experience of new materialism (unevenly recognized in the aforementioned critiques) to be of more potential use at this historical juncture than the dislocation and disembodiment of appeals to deep time.

7. See especially Jeffrey J. Cohen, *Stone: An Ecology of the Inhuman* (Minneapolis: University of Minnesota Press, 2015); Jane Bennett, *Vibrant Matter: A Political Ecology of Things* (Durham: Duke University Press, 2010).

8. See Noah Heringman, *Romantic Rocks, Aesthetic Geology* (Ithaca: Cornell University Press, 2004). Heringman relates this argument to the Anthropocene in chapter 3 of the present volume.

9. (2004), qtd. in Samuel Otter and Geoffrey Sanborn, "Introduction," in their *Melville and Aesthetics* (New York: Palgrave Macmillan, 2011), 1.

10. In classifying Hitchcock's period as Romantic rather than Victorian, I am referencing the fact that both geology and Romanticism operate according to different chronologies in the United States and in the United Kingdom. Although the scientific and literary names I listed at the outset of this chapter are likely more familiar to my readers than Hitchcock's, my focus on the latter's writing is in keeping with my desire, in the larger project from which this chapter is drawn, to expand the archive of authors recognized for speculative thinking about geology in the American nineteenth century beyond the small handful usually cited (Emerson, Thoreau, Melville, Dickinson).

11. See Edward Hitchcock, *Final Report on the Geology of Massachusetts* (Northampton: J. H. Butler, 1841).

12. See Ralph O'Connor, *The Earth on Show: Fossils and the Poetics of Popular Science* (Chicago: University of Chicago Press, 2007); Adelene Buckland, *Novel Science: Fiction and the Invention of Nineteenth-Century Geology* (Chicago: University of Chicago Press, 2013).

13. See Dana Luciano, "Sacred Theories of Earth: Matters of Spirit in William and Elizabeth Denton's *The Soul of Things*," *American Literature* 86.4 (2014): 713–36.

14. Qtd. in Rebecca Bedell, *The Anatomy of Nature: Geology and American Landscape Painting, 1825–75* (Princeton: Princeton University Press, 2002), 3.

15. Walt Whitman, "Slang in America," *North American Review* 141.344 (1885): 431–35, 435.

16. Qtd. in Bedell, *Anatomy of Nature*, 5.

17. See, e.g., C. F. Winslow, *The Preparation of the Earth for the Intellectual Races: A Lecture, Delivered at Sacramento, California, April 10, 1854, at the Invitation of the House of Assembly* (Boston: Crosby, Nichols, 1854).

18. Qtd. in Bedell, *Anatomy of Nature*, 5.

19. Gayatri Spivak, "Translator's Preface," in Jacques Derrida, *Of Grammatology* (Baltimore: Johns Hopkins University Press, 1976), xvii (hereafter cited parenthetically in text).

20. Steven Shaviro, "Specters of Marx," *Pinocchio Theory*, February 8, 2006, http://www.shaviro.com/Blog/?p=474.

21. Anthony J. Martin, *Dinosaurs Without Bones: Dinosaur Lives Revealed by Their Trace Fossils* (New York: Pegasus, 2014), 9.

22. Untitled headnote, *Knickerbocker* 8 (December 1836): 750.

23. Ibid.

24. Though scientific didactic poetry's classical Latinate form had begun to decline by the time geology took modern form, Anglophone adaptations flourished through the nineteenth century. See Robert M. Hazen, *The Poetry of Geology* (London: Allen Unwin, 1982).

25. For one such reading, see Jordan D. Marché II, "Edward Hitchcock's Poem, *The Sandstone Bird* (1836)," *Earth Sciences History* 10.1 (1991): 5–8.

26. Charles Babbage, *The Ninth Bridgewater Treatise: A Fragment*, 2nd ed. (London: John Murray, 1838) (hereafter cited parenthetically in text).

27. Tina Young Choi, "Natural History's Hypothetical Moments: Narratives of Contingency in Victorian Culture," *Victorian Studies* 51.2 (2009): 275–97, 283.

28. Edward Hitchcock, *Ichnology of New England* (Boston: W. White, 1858), 173 (hereafter cited parenthetically in text).

29. Serenella Iovino and Serpil Oppermann, "Introduction," in their *Material Ecocriticism* (Bloomington: Indiana University Press, 2014), 4.

30. See Bennett, *Vibrant Matter*.

31. Edward Hitchcock, *The Religion of Geology and Its Connected Sciences* (Boston: Phillips, Samson, 1857), 413 (hereafter cited parenthetically in text).

32. The quotation from Babbage's text is one of only two citations of the slave system in *The Religion of Geology*; the other is a fleeting mention of slavery as a manifestation

of evil. For his part, Hitchcock followed the extended quotation from Babbage by developing his speculations on the endless chain of impressibility. Returning to his earlier suggestion that a "higher sphere" might be able to discern such traces, he hypothesized that the traces might well affect that sphere as well, transcending the material realm to produce a further reaction, transmuted but still resonant in the celestial afterlife. The suggestion that temporal deeds did not simply determine what one's posthumous fate would be but continued to act upon the afterlife implicitly extended Babbage's contention still further, suggesting that the slave trade had done damage not only to the planet, but to eternity.

33. Wai Chee Dimock, *Empire for Liberty: Melville and the Poetics of Individualism* (Princeton: Princeton University Press, 1991).

34. Henry Wadsworth Longfellow, "To the Driving Cloud," in his *The Belfry of Bruges and Other Poems* (Cambridge: John Owen, 1845), 66–70.

35. Many thanks to Tavia Nyong'o for clarifying this point.

36. Caleb Smith, *The Oracle and the Curse: A Poetics of Justice from the Revolution to the Civil War* (Cambridge: Harvard University Press, 2013), 36.

37. Mark Seltzer, "The Official World," *Critical Inquiry* 37.4 (2011): 724–53, 729.

38. Bennett, *Vibrant Matter*, xiv.

39. New materialist scholarship is giving increasing attention to the imbrication of bodies and matter as a necessary component of social justice thought; see, for example, Stacy Alaimo, *Bodily Natures: Science, Environment, and the Material Self* (Bloomington: Indiana University Press, 2010); and Jussi Parikka, *A Geology of Media* (Minneapolis: University of Minnesota Press, 2015). This focus remains an uneven one in inhumanist thought, however, not only because the desire to, as Karen Barad puts it, "give matter its due as an active participant in the world's becoming" in the face of a history of anthropocentric thought has led to a consequent underattentiveness to the human, but also because of a concern about the potential "negativity" involved in uncovering histories of social damage as eclipsing the "positive and constructive" approach that the new materialisms, in particular, seek to foreground. Barad, *Meeting the Universe Halfway: Quantum Physics and the Entanglement of Matter and Meaning* (Durham: Duke University Press, 2007), 136. See also Bennett, *Vibrant Matter*; Diana Coole and Samantha Frost, "Introduction," in their *New Materialisms: Ontology, Agency, Politics* (Durham: Duke University Press, 2010), 9.

40. Simon Lewis and Mark Maslin, "Defining the Anthropocene," *Nature* 519 (March 2015): 171–80. As far as I know, no similarly geologically verified trace has been found of the Atlantic slave trade (which, as Sylvia Wynter notes, is inextricable from histories of European contact with the Americas—not a separate, concurrent action but part of the same event). Wynter, "Unsettling the Coloniality of Being/Power/Truth/Freedom: Towards the Human, After Man, Its Overrepresentation—An Argument," *CR: The New Centennial Review* 3.3 (2003): 257–337.

41. I have written about this proposal in Luciano, "The Inhuman Anthropocene," *Avidly: Los Angeles Review of Books*, March 22, 2015, http://avidly.lareviewofbooks.org/2015/03/22/the-inhuman-anthropocene. See also Steve Mentz, "Enter Anthropocene, Circa 1610," *Glasgow Review of Books*, September 27, 2015, http://glasgowreviewofbooks.com/2015/09/27/enter-anthropocene-c-1610.

42. Jan Zalasiewicz et al., "Colonization of the Americas, 'Little Ice Age' Climate, and Bomb-Produced Carbon: Their Role in Defining the Anthropocene," *Anthropocene Review* 2.2 (2015): 117–27, 127. See also Clive Hamilton, "Getting the Anthropocene So Wrong," *Anthropocene Review* 2.2 (August 2015): 102–7.

43. Wynter, "Unsettling the Coloniality."

44. I adapt this term from Lauren Berlant's *Cruel Optimism* (Durham: Duke University Press, 2011). As Berlant notes in the final chapter, the Anthropocene itself may be conceived in this frame. My disinclination to critical awe parallels Berlant's impatience with trauma as the affective sign of the present.

# 6.

## Partial Readings
### Thoreau's Studies as Natural History's Casualties

*Juliana Chow*

The scale on which his studies proceeded was so large as to require longevity, and we were the less prepared for his sudden disappearance.
—Ralph Waldo Emerson, "Thoreau"

Almost as if in answer to Emerson's view of the scale of Thoreau's work, studies on climate change have reassessed Thoreau's herbarium and phenological records and compared them against measurements more than 150 years after his death.[1] Beginning in 1860, Thoreau compiled phenomenal observations from his journal into charts that scholars now call the Kalendar. While some of these observations mark temperature-sensitive fluctuations of specific phenomena, like plant leafings and flowerings and bird migration, others mark Thoreau's own fluctuations across months—the cadence of his activities in response to seasonal shifts. In her careful perusal of Thoreau's Kalendar, Kristen Case remarks on the variety of objective and subjective kinds of observation, and acknowledges, "The charts of 'general phenomena' are equally resistant to both scientific and literary analysis. The entries cannot be converted to data points but neither can they be paraphrased, quoted, or analyzed in terms of something like literary 'style.'"[2] Yet it is precisely a kind of analysis at once scientific *and* literary that Thoreau's charts and late natural history writings prompt when researchers work with the references in these observations in order to measure anthropogenic environmental change.

For Thoreau, writing literature is, as Sharon Cameron puts it, "writing nature."[3] And writing nature is related to thinking, reading, and writing about historiographic and literary form. Following from Emersonian transcendentalism,

Thoreau understood natural history as meaningful in relation to human history, but it was through his regional studies of ecology and forest succession that Thoreau came to understand natural history as merged with human history. To take words from *Walden*, his work was as much figuring out how to "improve the nick of time" as it was simply marking time and change.[4] Considering Thoreau's literary perambulations from a posthumous perspective in which those perambulations ultimately cease and disappear, how might we read Thoreau's writings as ecological and epistemological engagements with abrupt cessation, with incompleteness and partiality? Thoreau's method offers a sense of coarticulated human and natural history—a partial view that palpably registers loss—as well as a partiality for small clearings and gatherings: woodlots to trespass instead of the woods in general, lacunae instead of panoramas of terrain, discrete lives instead of the totality of life.

Attending to his most intensive and incomplete study of forest ecology, "The Dispersion of Seeds" (1860–61), I look at how Thoreau situates his work in relation to geologic time as a way to read what he calls the "history of a woodlot."[5] Thoreau's late writings suggest to me a concurrence of biological, literary, and historical forms based on ecological relations of partialities rather than organic wholes, of dispersals rather than monologic continuity. I turn to "Dispersion" because it is a text in which Thoreau foregrounds ecology—he had just read Charles Darwin's *Origin of Species* (1859)—and because it is a manuscript draft whose formal difficulties raise the question of how to read partial forms *and* read from an orientation I am calling "critical partiality."

Critical partiality is an attempt to come to terms with how an approach may be critical at the same time that it is experimental and tentative. It is a mode of being partial, partial *to* something, partial *of* something, with partiality *as* form itself, and partiality as a potential form of history and of experience, as partial histories or partial knowings or not knowings. In her essay "Reluctant Ecology in Blake and Arendt," Amanda Jo Goldstein describes an "experimental knowledge" in Goethe that is "alive at once to the fragility of the object and to the vulnerability of the knower in the scene of experimental trial."[6] This is a relation between subject and object that is more a process of participating in or becoming permeated with each other, knowing as intimate and precarious.

*Part* of a process, *part*icipate, im*part*, a*part*—perhaps if I could enumerate all the kinds of partiality I could at last put my finger on what this is. Goldstein invokes a confluence with Donna Haraway's theory of feminist objectivity and situated knowledge, of a partial and embodied perspective: "And by 'partial'

I mean biased, incomplete, limited and compromised by contact, one way to understand 'figurative' as an alternative to 'false'" ("Reluctant Ecology," 152). How do we approach things as figures? How do we figure out something? How do we ourselves figure in something? How can a fact be taken as figural and partial, as figuration? Or, how do we understand reference and cross-reference to be, as C. S. Giscombe puts it, "an unqualified marker, some days the ache of an implicit region"?[7] Maybe a critical partiality remains attentive to and part of what Goldstein notices as fragility and vulnerability, to the provisionality of what I have arranged as a chapter on Thoreau's writings. Part of something, but barely. The contingency of our happening to be, our trying. What is the shape of vacancy? Not history as survey or totality or even as collective project, but history as casualty.

Ecology has, since Ernst Haeckel named the science in the mid-nineteenth century, been concerned with fitting together environmental relations into a whole, with investigating, in his words, "the total relations of the animal both to its inorganic and its organic environment. . . . [Ecology] is the study of all those complex interrelations referred to by Darwin as the conditions of the struggle for existence."[8] Announcing, on the one hand, theories of evolution as the provocation for ecology—that is, tying the concept of relations to the concept of succession—Haeckel specifies, on the other, a perspective of comprehensiveness and systemic relations. It is this ecology, the study of "total relations," that continues to be relevant for concepts of a human-defined epoch and environmental change on the scale of the whole Earth system.

As early as 1864, George Perkins Marsh's macrocosmic study *Man and Nature* takes up a modern ecological perspective and posits "geological agency" in organic life.[9] Cited by Paul Crutzen and Eugene Stoermer in their proposal for using the term "Anthropocene," Marsh's work may be seen as one genealogical antecedent to the concept.[10] Marsh warns of "changes produced by human action in the physical conditions of the globe we inhabit" through "operations which, on a large scale, interfere with the spontaneous arrangements of the organic or inorganic world" (*Man and Nature*, 3). Although he sees human action as an external destructive force, this is complemented by his belief in potential rectification, emphasizing human agency as an overarching regulatory power in ecological relations for better or worse.

Thoreau's sense of the discontinuities of life departs, however, from these modes of analysis that reinforce ecology as a theoretical and ontological totality and that, in the case of Marsh, give us a pulse of nineteenth-century physical geography that corresponds to the momentum of an Anthropocene concept. In the very moment when ecology was becoming a science, Thoreau's writings staged the issues of wholeness and partiality that theorists of anthropogenic environmental change now address. Dipesh Chakrabarty has argued that the Anthropocene is beyond human rationalization since it marks the transformation of the human species into a geological force and the merging of human history with natural history; and yet it is also an ongoing, lived process that unevenly affects populations and is unevenly attended to by humans.[11] Chakrabarty's multiscaled Anthropocene history signals the aporia of a historiography that cannot reconcile what he calls "nonhuman-human" agency, which in its collectivity becomes a "nonliving" force of nature, with "human-human" concerns for environmental justice ("Postcolonial Studies," 11). The nonhuman-human exists at the abstract level of humanity and geology taken together as a whole, while the human-human resides at the concrete level of human experience. This tension between a unified view of the Earth system and, say, a partial view of a biome may be interpreted as problems of scale and perspective, etic or emic observational standpoints, distance or closeness, generalization or specificity.[12]

Thoreau's "Dispersion of Seeds" presents a version of ecology that holds the tension between these views. Partly a continuation of his address "The Succession of Forest Trees" given in September 1860 at the Middlesex Cattle Show, "Dispersion" endeavors to show how pines succeed oaks and vice versa through various means of seed transport. While the history of woodlots remains the framework, Thoreau's manuscript also traces seed transport in other plant species, including common weeds of New England forests. Thoreau's text insists on the distinct processes of each species so that the transport of pine cones and acorns does not stand in for the seed dispersal of all. Rather than relinquishing a partiality to the whole, "Dispersion" allows for discontinuity and the *parts* of relation in its accounts of succession. This attention to life's dispersals illuminates how we might think of the exigencies and differentials of succession in the Anthropocene era through partial and dispersed forms held in their caesura rather than absorbed into the whole. And in caesura—in the form and shape that caesura takes—we might know not only how such partial forms figure in life, but also how language figures in ecology.

In the kernel of "Dispersion" that focuses on the history of woodlots, Thoreau's interest in what he calls "forest geometry" becomes an occasion for comparing his method to that of geologists (166). Indeed, the geometric spread of species literalizes temporally organized relationships based on breaking off and trespassing. Studying how a forest or a plant species expands and contracts and covers the surface of the Earth in shapes, Thoreau does not simply document ecological relations, but also predicts environmental shifts based on those relations. In other words, his understanding of the ecology underlying changing forest geometries marks the point at which diachronous concepts of change, like Darwin's theory of evolution, interrupt (though eventually become interpolated into) the synchronization that a whole is based on. Thoreau writes of such contingent processes, which transpire slowly over long periods of time, "In this haphazard manner Nature surely creates you a forest at last, though as if it were the last thing she were thinking of. By seemingly feeble and stealthy steps—by a geologic pace—she gets over the greatest distances and accomplishes her greatest results" (36). In conceiving forest evolution at a geologic pace, Thoreau signals his interest in Lyellian uniformitarianism, but also Darwinian developmental change. However, Thoreau retracts this analogy in subsequent paragraphs. Geologic pacing turns out to be an inaccurate universal statement when one is dealing with the more rapid and nonsynchronous living periodicities of pines versus oaks, beds of cress and radishes versus trees. These are not the "feeble and stealthy steps" of one nature, but the steps of many species meeting each other at different tangents, overlaps, rates, periods, and coincidences. As Thoreau observes, oaks do not miraculously succeed a pinewood but depend on a series of ecological relations: acorns are transported by squirrels to pinewoods, and oak seedlings depend on the pines' shelter to grow; once the pines are cut down and harvested, the oaks naturally succeed.

Thoreau makes crucial distinctions between his protoecological study with its multifaceted timings and the study of strata as a monolithic record of the Earth's history. At one point in his discussion, Thoreau notes that "geologists tell us that the coniferous trees are older, as they are lower in the order of development, than oaks" ("Dispersion," 168). Paragraphs later, distinguishing earlier successions of wood based on stumps and their rings, he muses, "in this case we have one advantage over the geologist, for we can not only detect the *order* of events but often the time during which they *elapsed* by counting the rings on the stumps. Thus you can unroll the rotten papyrus on which the history of the Concord forest is written" (169, emphasis added). What is worthwhile in a papyrus history of woodlots as opposed to a fossil record of tree species? Thoreau's attention to elapsing time, along with succession, emphasizes the decaying of life that unravels as you read it. While some mid-nineteenth-century geologists were motivated to organize the taxonomic order of more primitive versus more modern species by dating their appearance in the fossil record, a plant reader like Thoreau wanted to see something more than order since time is not conceived only along one timeline or stratification, in which species appear in sequence. Rather, there is also the time of each tree in relation to other trees, each making their own layers of years, each recording a curling of time, so that the living—and dying—periodicities of each tree may be placed in relation to the periodicities of other trees, creatures, seasons, and environmental fluctuations. Reading the durations of these discrete lives in concurrence with one another allows Thoreau to make inferences about ecological relations or "geometries" in which one species' life and death is contingent upon another's, as the oak is to the pine is to the farmer. For Thoreau, keeping count of elapsing time is simultaneously keeping track of species lapsing from life to death in relation to one another.

In "Dispersion," Thoreau's portrayal of ecological relations may be an intensely detailed "history of the Concord forest," but that history is intertwined with a human-natural history perceived at a scale that encompasses the whole Earth system. Thoreau's research in seed transport leads to his sense of a transoceanic ecology. Charting the traversals of common weeds, including dandelions, daisies, thistles, and fireweed, Thoreau is well aware of the transatlantic commerce and cultivation of plants that has led to weeds flourishing in new areas and along routes of passage and the problem of dominating species, like the nonnative Canada thistle but also native pines and birches, which threaten to homogenize an ecosystem. He reports on extraordinary accounts of one plant's dispersal: "A late American species of the same genus (*Erigeron canadensis*) has become a common weed in Europe and is found, according to De Candolle, as far as Kasan. Mrs. Lincoln says that 'Linnaeus asserted that the *Erigeron canadensis* was introduced into Europe from America by seeds wafted across the Atlantic Ocean.' But of course, they could not wait for Columbus to show them the way" (83–84). Though Thoreau makes light of Columbus's role in this process, some scientists now consider the homogenization of species biodiversity (and mass extinctions and genocide) precipitated by New World colonization to be among the order of environmental shifts visible as a stratigraphic marker.[13] In imagining seed vessels gathering at the edges of new shores, Thoreau references the mutual histories—human and natural—that overlap and depend on one another at a global ecological scale, crossing oceans and continents.

Why, then, does he remain fascinated with his own empirical and archival researches of New England plant species? In Thoreau's protoecology, the object of study is not the appearance of species in a geographic locale, but the mechanism of their transport. Assessing the feasibility of Linnaeus's claim, Thoreau makes his own observations of thistledown coasting above Walden Pond:

> [A]t five o'clock one afternoon last year just after rain, being on the middle of Walden, I saw many seedless thistledowns (sometimes they are seeded) sailing about a foot above the surface; yet there was little or no wind. It is as if they were attracted to the pond and there were a current just above the surface which commonly prevented their falling or rising while it drove them along. They are probably wafted to the water from the neighboring hollows and hillsides where they grow, because the currents of air tend to the opening above the water as their playground.
> 
> Here is a wise balloonist for you, crossing its Atlantic—perhaps going to plant a thistle seed on the other side; and if it comes down in a wilderness, it will be at home there. ("Dispersion," 85)

Through the miraculous buoyancy of thistledown, seeds travel by floating upon the strange dynamics of air in the clearings created by bodies of water. Against the nineteenth-century commonplace that some plants seem to spontaneously generate, Thoreau affirms that plants come from seeds and that seeds come from somewhere. But through the metaphor of seed transport as ballooning, he also aligns the mechanism with a certain whimsy that is corroborated in one nineteenth-century commentator's verdict that ballooning is "a sublime but profitless philosophic toy."[14] The playfulness of seed transport as a balloon full of hot air matches Thoreau's careful note that mostly there is no seed in this thistledown. Yet the transport mechanism is in place even if it is empty, and this preserves the improbable possibility of a seed making its way across the ocean. Because it is the means of transport and not the seed that actually garners Thoreau's attention, the ballooning thistledown suggests that the metaphorical aspects of this mechanism are crucial in his thinking about ecological relations and succession. Transport, or metaphor, is a mechanism that can ensure continuity of movement without necessarily continuity of content.

Thoreau's metaphor of seed transport is, moreover, a metaphor of differentiation rather than of representativeness. In recounting the contingencies of every seed's transport, "Dispersion" emphasizes both its own and its subject's dispersal. I turn to Thoreau's presentation of milkweed, *Asclepias cornuti*, as a set of instances:

> [T]he extremities of the silken threads detach themselves from the core and from being conduits of nutriment to the seed, perchance to become the buoyant balloon which, like some spiders' webs, bears the seeds to new and distant fields. (91)

> Perchance at the tops of some more open and drier pods is already a little flock of these loosened seeds and down, held by the converging tips of the down like meridians—just ready to float away when the wind rises. (91)

> The few seeds which I release soon come to earth, but probably if they waited for a stronger wind they would be carried far.
> Others, again if you wait a while, are found open and empty, except of the brown core, and you may see what a delicate, smooth, white- or straw-colored lining this casket has.
> If you sit at an open attic window toward the end of September, you will see many a milkweed down go sailing by on a level with you, though commonly it has lost its freight—notwithstanding that you may not know of any of these plants growing in your neighborhood. (92)

> When I release some seeds, the fine silky threads fly apart at once, opening with a spring—and then ray their relics out into a hemispherical form, each thread freeing itself from its neighbor, and all reflecting prismatic tints. . . . I let one go . . . and I fear it will make shipwreck against the neighboring wood. But no. (92)

> I brought home two of these pods which were already bursting and amused myself from day to day with releasing the seeds and watching them rise slowly into the heavens till they were lost to my eye. (93)

The drift of milkweed streamers over the neighboring wood and of thistledown over Walden Pond are not synecdoches for transatlantic species migration. But each instance remains both a metaphorical and a material part of the vast process of species migration, "its Atlantic" qualified by a more limited possession. The iterative quality of the milkweeds in "Dispersion" and their differences from each other, which Thoreau conveys through descriptions of the varieties of transport, are not reducible to a singular instance from which we may generalize the experience of the whole into one "telling" metaphor. Rather than a synecdoche—a taking up together—Thoreau's metaphor works in the opposite way: as a *merismus*, or dividing, where parts might add up to a whole but remain distinct parts: thread, pod, tip, casket, ray; milkweed, milkweed, milkweed, milkweed. By Thoreau's articulation of so many instances of transport, the differentials of seed dispersal remain differentials rather than becoming representative aggregations.

Further, what rivets Thoreau in this manuscript is not making the kind of reference that endures, preserved in stratigraphic and taxonomic levels, but instead tracing the traversals that elapse, lapse, and slip across different orders of biological *and* textual forms. With the milkweed, Thoreau's "perchances" and "ifs" illustrate the slippages of seed transport, from the many fraught contingencies of conduit and landing to the annulled vessel that has lost its precious seed freight. Yet this hazardous dispersal is the very means by which life succeeds. Recalling both the spider's flight on its loosened web and shipwreck imagery, Thoreau's many instances of milkweed transport allow the haphazard venture of life's succession to be seen *through* and understood *as* dispersal rather than as undermined or counterbalanced by such a process.[15] Staging these flights of buoyant milkweed streamers—and of text—becomes an expression of the probabilities of a seed's and a trope's traversal succeeding.[16] The compilation of instances shows Thoreau's compositional method of gathering together journal entries on the milkweed pod bursting or cracking and releasing its seeds in downy airborne flocks. One might read this compilation as a relic of the manuscript's untrimmed state, but I imagine this form of haphazardness as written into Thoreau's text, which performs an openness to chance through these stagings. He concludes of the cloudlike milkweed: "At any rate, I am interested in the fate or success of every such venture which the autumn sends forth. And for this end these silken streamers have been perfecting themselves all summer, snugly packed in this light chest, a perfect adaptation to this end—a prophecy not only of the fall, but of future springs. Who could believe in the prophecies of Daniel or of Miller that the world would end this summer, while one milkweed with faith matured its seeds?" ("Dispersion," 93).

The anomalous miracle in dispersal's seemingly self-destructive enterprise turns out to be a "perfect adaptation" for the outlier of transplanted seed. Through merismus, Thoreau's partial view appears to proleptically reject the Anthropocene's urgently apocalyptic narratives even as he is subtly aware of the environmental devastations that transoceanic traversals catalyze and accrete.

If the Anthropocene fuses together human and natural history so seamlessly that only mutual destruction or restoration is possible, Thoreau's partial ecology is less willing to marry parties into a whole as it literally lets ephemera slip and disperse.

Reading how Thoreau engages with a nascent ecological science in "The Dispersion of Seeds" has also come to mean reading the text as a mechanism of that science, an aesthetic technique of staging one's experimental knowing. At the heart of such an attempt is to read (and write) literary form in relation to ecological and historiographic form. In other words, how might we read a text (or a fossil record) if we understand the text or a figure to be partial rather than whole? To be apart or dispersed in merismus rather than continuous with or metonymic of a whole universe? Where a part is not part of a whole, where part and whole do not get resolved? What happens if a reference cannot be traced because it gets lost, its tracks ending irretrievably?[17] In the posthumously published essay "Walking" (1862), Thoreau observes of the pine trees on Spaulding's Farm: "They fade irrevocably out of my mind even now while I speak and endeavor to recall them, and recollect myself. It is only after a long and serious effort to recollect my best thoughts that I become again aware of their cohabitancy."[18] If a reader traces a referent—pine tree, milkweed, thistledown—across overlapping fields of meaning in literature and the sciences, assuming one can see its transports and organize them in an essay for synoptic perusal, that reader might pause at the transformations to account for what is lost. This is transport not as preservation or succession of a continuity, but transport as loosing and losing something. This is a long pause at the boundaries of crossings.

Thoreau's own lapse might be marked in this way: in writing and thinking of his sketches, Thoreau held the tension between form as discontinuous and unsynthesizable and form as a unified whole. Musing on why he left Walden more than four years previously—"But Why I changed—? Why I left the woods? I do not think I can tell"—Thoreau struggles to account for Walden as an interval in his life in order to, in his words, "set down such choice experiences that my own writings may inspire me and at last I may make wholes of parts."[19] Yet a few days later on January 27, 1852, he reverses this aim of writing a whole in favor of both a different frame and a different sense of words: "I do not know but thoughts written down thus in a journal might be printed in the same form with greater advantage—than if the related ones were brought together into separate essays. They are now allied to life—& are seen by the reader not to be far fetched—It is more simple—less artful—I feel that in the other case I should have no proper frame for my sketches. Mere facts & names & dates communicate more than we suspect—" (*Journal*, 3:239). If Thoreau attempts to make wholes out of parts by extracting and revising from his journal entries, he also begins to consider whether the parts are more "whole" and alive in the journal itself.

This sequence has been read to posit Thoreau's journal as a whole text and thus a literary work as much as *Walden* is. However, in requiring a proper frame, Thoreau requires separation between parts, whether they are the duration of time between experience and writing, or infills of text between the journal entries he might extract for essays, or lines drawn to form a chart organizing his observations of phenomena by year. Through this partiality for framings—which are also lapses—Thoreau does not ascribe "wholeness" to his journal but a certain quality of relation that relies on separation. Somehow, the journal's parts, and even his jotted notes of facts, names, and dates, convey this relation as being not "far fetched" but proximate and aligned to the point of cleaving together.

Thoreau implies, then, a relation of text to world that is not so much representation or index, but relation itself—not to *seem* or to *be* life, but to be "allied to life." If text is merely in relation to life, and transport is prone to falter and formally dissolve, this alliance is a tenuous one. Yet it is this very tenuousness that allows Thoreau to revise and rearrange his materials in multiple formats. Through their textual history, Thoreau's works attest to a form of writing, and of ecology, as a dispersal into the journal, *Walden*, essays, manuscripts, charts, and notes; *and* they attest to the intervals and lapses between text and life.

In "Walking," Thoreau gives voice to these uneasy intervals as a question of not only formal but also epistemological partiality. Contrasting his surveys of woodlots against his forays, he finds himself at a loss in between:

> For my part, I feel, that with regard to Nature, I live a sort of border life, on the confines of a world, into which I make occasional and transient forays only. . . . These farms which I have myself surveyed, these which I have set up appear dimly still as though through a mist; but they have no chemistry to fix them; they fade from the surface of the glass; and the picture which the painter painted stands out dimly from beneath. The world with which we are commonly acquainted leaves no trace, and it will have no anniversary. (217–18)

The viewing glass to compose one's view for landscape description is the picturesque frame that Thoreau obliquely occupies in his natural history writings. It is a mixed metaphor that draws upon painterly gestures and photographic processes to figure his mensuration and mapping work as a surveyor. As usual with such inconsistent metaphors, which are, in effect, a series of partial metaphors that slip from one to the next, each figure cannot hold the view: the exposures unfix, the pictures submerge, the boundaries dim. This dissolution of scientific and aesthetic conventions, as well as of the "common" world, does not afford clarity. There is no access to worlds, natural or conventional. Instead, the surveys, which are Thoreau's occasions for trespassing into the woods, are forays that do not hold. The epistemological imperative to survey stalls and mists—it transpires in partial readings, partial figurations.

Toward the end of "The Dispersion of Seeds," Thoreau relates an anecdote of a cow who strayed into his front yard. Scratching against and butting her head into a tree he had just planted, the cow broke the young sapling mid-stalk, and Thoreau observes the aftermath:

> However, it [the sapling] survived this accident, and the numerous low boughs which spread over the ground gradually erected themselves around the center, so that instead of one slender spring stem, I have now a dozen forming one dense tree of a beautiful and perfectly regular conical form. A neighbor who has some of these trees trimmed off in the usual stiff manner, which did not content him, lately took the pains to come to me and ask how I had treated my arbor-vitae to produce such a result. I told him that all he had to do was to leave his gate open at the season when cattle are driven up-country. (170–71)

Thoreau's arborvitae recalls Darwin's ramifying tree of life but one that is not just nature—a tree that is both nature and art, cultivated *and* let to grow as it will. In this parabolic tale, we might read the cow as the denouement of the anthropogenic interferences of agriculture, pasture, and cattle, which disastrously cripple natural production, although it recovers. Or, we might read the cow as following some wild instinct or attraction of her own, as does the tree, and the gate left open as being open to the fortuitous accidents and adaptations that may occur. How do we manage chance or leave room for accident? Or, how do we let go and let slip what has been taken to be proper ratios and shapes, only to find them returned to us as if something that took form before thought?

Thoreau's radical openness—which is also a form of lapse—is precisely to clear a way for such happenstances, to look for the transport even in wreck or void. This is not to say that openness is complacent. He calls for, after all, the appointment of forest wardens in the last line of "Dispersion." Although he criticizes farmers for having "never considered what is to be the future history of what they call their woodlots" (129), he also imagines that future history through potent, if partial, thought—a thought that depends on others to take it up in their partial way: "Thus, one would say that our oak forests, vast and indispensable as they are, were produced by accident, that is, by the failure of animals to reap the fruits of their labors. Yet who shall say that they have not a dim knowledge of the value of their labors?—that the squirrel when it plants an acorn, and the jay when it lets one slip from under its foot, has not sometimes a transient thought for its posterity, which at least consoles it for its loss?" (130). In this way, "loss" becomes a consideration of history, a loss that is not filled in, but consoled and made more solid, made into a shape with depth and edges and lips, with a different epistemic orientation—"a dim knowledge," "a transient thought." Through this, Thoreau proposes a way to read the history of woodlots, that is, a way to read human-natural history through a partial ecology rather than a total system. What is indicated, then, is how to read what is not preserved in the geological record or in any concept of systemic continuity—the casualties of a totalizing perspective. Thoreau's late ecological studies in "Dispersion," his phenomenological charts, and other writings might be understood as part of documentary ephemera drifting out toward an uncertain future and mixing with the exposed transports of various seeds, not collated and gathered, but dispersed, discrete, and potentially barren. In the parabolic openness with which Thoreau leaves us, in this immensity of loss, we may find ourselves sounding the depths: "They fade irrevocably."

In examining Thoreau's late writings, I imagine a critical approach that does not attempt a survey, that does not, in other words, follow an assumption of analytical or metonymic comprehensiveness, but holds that view in tension with a view of discontinuity and partiality. We might still know anthropogenic environmental change, not through a global or nonhuman structure of thinking, but through regional and experiential techniques of presentation. Further, although Thoreau's manuscripts, charts, and notes might appear analytically unviable or incomplete, they register the costs of systemic thinking, where a certain kind of natural history itself is a casualty of an epistemic formation. This costly move, from Thoreau's natural history to an ecology of "total relations," from finite and discrete lives to a monologic continuity of life, troubles our reading of the Anthropocene.

# Notes

1. The Primack Lab of Boston University has produced data for publications in a variety of genres (from academic reports to popular science writing) and disciplines that make use of Thoreau's records for the study of climate change. See, for example, Charles G. Willis et al., "Phylogenetic Patterns of Species Loss in Thoreau's Woods Are Driven by Climate Change," *Proceedings of the National Academy of Sciences* 105.44 (2008): 17029–33; Richard Primack and Abraham Miller-Rushing, "Uncovering, Collecting, and Analyzing Records to Investigate the Ecological Impacts of Climate Change: A Template from Thoreau's Concord," *BioScience* 62.2 (2012): 170–81; and Richard Primack, *Walden Warming: Climate Change Comes to Thoreau's Woods* (Chicago: University of Chicago Press, 2014).

2. Kristen Case, "Knowing as Neighboring: Approaching Thoreau's Kalendar," *J19: The Journal of Nineteenth-Century Americanists* 2.1 (2014): 107–29, 123.

3. In *Writing Nature* (New York: Oxford University Press, 1985), Sharon Cameron argues more specifically that Thoreau's journal is *his* form of writing nature, expressing a wholeness of nature that is, however, "predicated not on connections but on the breaking of connections" (6). My argument in this chapter takes inspiration from hers, but I consider his late writings in particular, where critics and historians have noted his increasingly scientific attention to natural history as part of this writing nature.

4. Henry David Thoreau, *Walden*, ed. J. Lyndon Shanley (Princeton: Princeton University Press, 1971), 17.

5. Henry David Thoreau, "The Dispersion of Seeds," in *Faith in a Seed: The Dispersion of Seeds and Other Late Natural History Writings*, ed. Bradley P. Dean (Washington, D.C.: Island, 1993), 170 (hereafter cited parenthetically in text).

6. Amanda Jo Goldstein, "Reluctant Ecology in Blake and Arendt: A Response to Robert Mitchell and Richard Sha," *Wordsworth Circle* 46.3 (2015): 143–55, 145 (hereafter cited parenthetically in text).

7. C. S. Giscombe, *Prairie Style* (Champaign, Ill.: Dalkey Archive Press, 2008), 19.

8. Qtd. in Carolyn Merchant, *The Columbia Guide to American Environmental History* (New York: Columbia University Press, 2002), 160.

9. George Perkins Marsh, *Man and Nature*, ed. David Lowenthal (1864; repr., Cambridge: Harvard University Press, 1967), 69 (hereafter cited parenthetically in text). Marsh's work follows in the tradition of physical geographers like Alexander von Humboldt whose *Cosmos* (1845, 1847; translated into English in 1849) attempted an inductive study of empirical data from tiny organisms to the planets of outer space, and from the material world to the intellectual and cultural realm. Marsh's scientific sources range widely: continentally, disciplinarily, and anecdotally. His is a study insistent on extensive and detailed *coverage*. In *The Passage to Cosmos: Alexander von Humboldt and the Shaping of America* (Chicago: University of Chicago Press, 2009), Laura Dassow Walls remarks that Marsh was "the most profoundly Humboldtian of all Humboldt's American children" (295), though he also modified Humboldt for a Christian doctrine that emphasizes human exceptionality and moral responsibility.

10. Paul Crutzen and Eugene Stoermer, "The 'Anthropocene,'" *Global Change Newsletter* 41 (May 2000): 17–18.

11. Dipesh Chakrabarty, "Postcolonial Studies and the Challenge of Climate Change," *New Literary History* 43.1 (2012): 1–18 (hereafter cited parenthetically in text).

12. I borrow the distinctions between unified or "external" and partial or "internal" approaches from Barbara Herrnstein Smith, who helpfully characterizes these differences as etic and emic and as both between and within the humanities and sciences as disciplines or "cultures." Smith, *Scandalous Knowledge* (Durham: Duke University Press, 2005).

13. See Simon Lewis and Mark Maslin,

"Defining the Anthropocene," *Nature* 519 (March 2015): 171–80.

14. E. W. B. C., "On Ballooning," *Knickerbocker* 45 (February 1855): 154.

15. Thoreau's comparison of seed transport to spider webs has immediate intertextual and intratextual references to Pliny's recollection of the willow's cottony seed fluff as *in araneam abit* ("Dispersion," 58); these natural phenomena also recall Jonathan Edwards's "'The Spider Letter," in *The Works of Jonathan Edwards*, ed. John E. Smith et al. (New Haven: Yale University Press, 1995), 1–8. Thoreau's understanding of the spider's flight and seed dispersal as a means of succession, albeit precarious, revises Edwards's musings on how the spider's flight to ocean burial signifies a divine act of mass destruction to achieve an ecological balance against an otherwise swarming and overpopulous insect life.

16. Thoreau's form of probability may also be contrasted against Emerson's transcendental statistics. See especially his essay "Fate," in Emerson, *Essays and Lectures* (New York: Library of America, 1983), 941–68. Where Emerson relies on the average tendency, Thoreau's interest in discontinuity tends to the outlier. In *Uncertain Chances* (Oxford: Oxford University Press, 2012), Maurice Lee has examined the rise of probabilistic reasoning in the nineteenth-century United States and differently argues that Thoreau's embrace of statistical averages paired with the instability of chance led him to reach for the approximation of natural laws through a narrow and limited, rather than holistic, empiricism. Further, Lee de-emphasizes Thoreau's literary aesthetics in relation to his scientific philosophy, pragmatically concluding that "Thoreau finds statistical reasoning aesthetically stimulating" (143). I would like to consider Thoreau's textual forms as critical (or critically incidental) parts of his engagement with ecology.

17. I allude here to Barbara Johnson's deconstructionist reading in *A World of Difference* (Baltimore: Johns Hopkins University Press, 1987) of Thoreau's hound, bay horse, and dove in *Walden* as catachrestic signs for loss itself.

18. Thoreau, "Walking," in *Excursions*, ed. Joseph J. Moldenhauer (Princeton: Princeton University Press, 2007), 219 (hereafter cited parenthetically in text).

19. Thoreau, *The Journal of Henry David Thoreau*, ed. Bradford Torrey and Francis H. Allen (New York: Dover, 1962), 3:217 (hereafter cited parenthetically in text).

# 7.

# Scale as Form

## Thomas Hardy's Rocks and Stars

*Benjamin Morgan*

Concepts of scale have become central to critical discourse about the Anthropocene both because the Anthropocene aspires to name a new scale of human agency and because the discussion asks us to resituate historical consciousness in relation to an expanded scale of geological time. The question of how one might reconcile or move between vastly divergent scales—by no means guaranteed—is a central matter of concern. Responding to the climate crisis is challenging, according to Dipesh Chakrabarty, because it requires us to think across the "rifts" between human scales of ethics and politics and nonhuman scales of deep time.[1] Mary Stiner and colleagues argue that we ought to understand the Anthropocene narrative in the context of a long history of scalar leaps in population and energy use that extends deep into prehistory; and Julia Adeney Thomas argues that the Anthropocene requires historians to scale their conception of the human both up to the level of the species and down to the level of the microbiome.[2] Making sense of the climate crisis, in all of these cases, requires significant engagement with problems of scalar multiplicity and incommensurability.

This chapter explores the possibility that literary scholars may enter this dialogue about scale most fruitfully by embracing a return to formalism. This proposition may initially seem counterintuitive: formalism is often defined against historically or politically invested modes of reading, and problems of scale are, most centrally, problems of history and politics. But a formalist will understand scalar leaps and disjunctures not only as facts but also as forms—forms that are subject to critical strategies for reading mediations, images, and narratives. To propose form as a point of contact between geology and literature is not new; the resonances between interpreting literary forms and reading geological formations have been recognized at least since the nineteenth century. Taking these resonances as the inspiration for a critical practice, I examine how formal aspects of two Thomas Hardy novels, *A Pair of Blue Eyes* (1873) and *Two on a Tower* (1882), navigate problems of scalar multiplicity and incommensurability. Hardy's narratives, informed by contemplations of deep time and deep space within Victorian geology, biology, and astronomy, are often structured by a disjuncture between human measurements of value and nonhuman magnitudes that expose all human life as infinitesimal. Narrative shifts in duration and perspective are among Hardy's techniques for navigating this scalar disjuncture. But against the notion that aesthetic form successfully reconciles human and nonhuman scales, I argue that the more important effect of Hardy's engagement with scale is to symptomatize a persistent failure to shift scales. The difficulty encountered by Hardy represents one aspect of a more fundamental problem of bringing nonhuman scales into the domain of the senses.

**Geoformalism**

To scholars of nineteenth-century science and literature, the Anthropocene, with its dramatic and disturbing narrative about the place of humanity in the sweep of geological time, might look like a throwback to a much earlier moment of Earth science. At its eighteenth- and nineteenth-century disciplinary origins, geology, as it drafted new stories about slow (uniformitarian) or sudden (catastrophist) change over time, was closely connected to narrative forms and genres. "The sciences of the earth became historical by borrowing ideas, concepts, and methods from human historiography," observes Martin Rudwick in one of two large books on the topic; Ralph O'Connor argues that spectacular narratives about Earth history were designed "to titillate readers as much as to instruct them."[3] Writing about this intersection of geology and nineteenth-century narrative, Adelene Buckland identifies "form" as a key term; this is not only because "geologists elaborated new *literary* forms" for narrating the events of Earth history, but also because form itself provided a point of contact between the "structures of forms of the natural world" and narrative forms, such as romance, epic, and realism.[4]

Even as geologists shifted from writing cosmological narratives to more sober projects of classification and dating, literary tropes persisted. In *A Treatise on Geology* (1837), John Phillips, the creator of one of the first geologic time scales, proposes that a scale of time is literally inscribed in rocks and remains there to be read: "There is, however, at any place considered alone, a scale left us in the crust of the earth, by which to measure exactly the order of antiquity among the terms of the series of organic life, and to compare the relative antiquity of these terms at different and remote places. . . . This scale is the series of stratified rocks."[5] Phillips aspires "to translate, as it were, the symbolical notation of the geological scale of time into intelligible periods" (*Treatise*, 1:10). Reflecting on the discipline's history in 1913, Arthur Holmes, a geologist who helped develop radiometric dating, observes that for earlier geologists, "the record [of stratified rocks] was perplexingly difficult to read, and the time units, based on estimates of present rates of erosion and sedimentation, were little more than guesses."[6] "Form" in early geology implies both geological and narrative structure; the material form of the Earth's strata is itself an origin story.

But the story told by strata posed an important challenge to the imagination in the nineteenth century: although the strata were clearly legible signifiers, reading them required nearly impossible feats of imagining inhuman time scales. Few were more attuned to this difficulty than Charles Darwin, whose theory of natural selection depended on an Earth history on the order of at least hundreds of millions of years. Darwin emphasized gaps in the rock record and the difficulty of imagining vast scales of time: the facts of geology cause "the mind feebly to comprehend the lapse of time."[7] Darwin recommended practices of direct observation that went beyond merely reading Lyell's *Principles of Geology*, which could only "give an inadequate idea of the duration of each formation or even each stratum"; the reader would also need to "for years examine for himself great piles of superimposed strata, and watch the sea at work grinding down old rocks and making fresh sediment, before he can hope to comprehend anything of the lapse of time, the monuments of which we see around us" (*Origin of Species*, 208). But before you devote years to pondering erosion, you might want to know that the task is doomed in advance: "the lofty pile of sedimentary rocks in Britain, gives but an inadequate idea of the time which has elapsed during their accumulation"; this spans "an infinite number of generations, which the mind cannot grasp" (210). In Darwin's view, adequately reading geological formations and strata would require an impossible durational awareness, beyond the reach of human faculties.

A brief survey of twenty-first-century popular and introductory writing in geology suggests that these metaphors of material legibility and tropes of cognitive failure have persisted in forums where the discipline is presented to novices or the public. Felix Gradstein and James Ogg, who coordinated the 2012 update to the official geologic time scale (GTS), explain in an encyclopedia entry that the construction of geological time involves "reading, organizing, and sorting Earth's stone calendar pages, and, as best as we can, reconstructing the content of any missing pages."[8] An undergraduate geology textbook begins by noting the same incommensurability between "our life experience" and geological time that had occurred to Darwin: "Once we start dealing with thousands of years our concept of the passage of these amounts of time becomes increasingly divorced from our life experience. . . . The geologist therefore has to develop a peculiar sense of time, and may consider 100,000 years as a 'short' period, even though it is unimaginably long when compared with our everyday life."[9] Popular books on Earth history adopt various strategies, often literary, for representing and conveying these vast time scales. Jan Zalasiewicz, the convener of the Anthropocene Working Group, opens *The Goldilocks Planet: The Four Billion Year Story of Earth's Climate* by explaining the chart of the GTS as a visual heuristic that might prevent readers from "becom[ing] hopelessly lost within such a quite inhuman timespan"; and in *The Earth After Us: What Legacy Will Humans Leave in the Rocks?* he adopts the science-fictional conceit of imagining "the Earth, in a post-human future, many millions of years hence, being re-explored" by aliens or some future form of life.[10] Geology, as a science of structural observation and of change over time, historically opens itself to the literary domain in two long-standing ways: through a disciplinary investment in tropes of legibility and inscription; and through a disposition toward temporal narrative forms, such as uniformitarianism and catastrophism, that attempt to bring geological time within the bounds of the human imagination.

A twenty-first-century revitalization of formalism as a literary-critical practice may allow us to think in a new way about both of these interfaces of the literary and the geological, since formalism has drawn inspiration from materialist theory and sought to disassociate the study of form from apolitical aestheticism. Drawing on the posthumanist orientations of object-oriented ontology, Sandra Macpherson argues for a conception of formalism that contests an anthropocentric association of form with the aesthetic by emphasizing form as the shape of a thing. Against Walter Benn Michaels's treatment of representation and form as synonymous—which excludes from the category of "form" natural entities like "rocks, and stones, and trees"—Macpherson argues for a conception of form that, nodding to Aristotelian hylomorphism, derives from shaped matter.[11] Contra Michaels, nonhuman nature also ought to be understood as having form: form is "nothing more—and nothing less—than the shape matter (whether a poem or a tree) takes."[12] Agency is important in that sentence—note that matter itself does the taking, without, necessarily, the intervention of a person who intentionally shapes it. But if Macpherson emphasizes that a materialist conception of form as shape creates an alternative to a position in which formalism must justify itself by laying claim to an extraformal domain (politics and history are the usual candidates), Caroline Levine inverts this logic, arguing instead that a capacious conception of form as shape or arrangement recuperates formalism as, precisely, a politically engaged practice. Form is *"an arrangement of elements—an ordering, patterning, or shaping,"* which means that formalism should be understood not as a commitment to pure readerly delight in aesthetic complexity, but rather as a study of practices of ordering (whether of syllables or human bodies) that joins aesthetics to politics.[13] In the background of Levine's reasoning are both Jacques Rancière and Theodor Adorno, the latter of whom grounds W. J. T. Mitchell's similar argument that formalism is always already "committed" by way of its "reflection on the precise way things are said or shown."[14] Materiality and arrangement—including political arrangements—are key in these accounts of formalism.

It is notable that stones and rocks often feature as a limit case in literary scholars' and philosophers' discussions of form. Michaels and Macpherson circle around Richard Rorty's claim that "anything—a wave pattern, an arrangement of stars, the spots on a rock—can be treated . . . as language."[15] One reason that stones are of particular interest in discussions of form is elucidated by George Eliot who, as she defines form in her brief, evocative essay "Notes on Form in Art," reflects on a distinction between human and geological forms: "what is form but the limit of that difference by which we discriminate one object from another?—a limit determined partly by the intrinsic relations or composition of the object, & partly by the extrinsic action of other bodies upon it. This is true whether the object is a rock or a man."[16] Eliot uses this distinction between intrinsic and extrinsic determination to propose a hierarchy of forms. If you split a stone in half, you have two stones, but the same is not true of a human. This means that the human form is higher than the form of the rock ("It is wholeness not merely of mass but of strict & manifold dependence")—and an artistic form, such as poetry, is more like a human than a rock ("Notes on Form," 435). The form of the rock is determined extrinsically by my splitting it in half (or, more often, by the eroding agencies of wind and water), but the form of my own person is defined intrinsically as the composition and internal relations of my body. What I would emphasize in Eliot's discussion of form, however, is not the distinction but the continuity: aesthetic forms arise from a conception of natural forms (Eliot's next example is a shell), and their differences connect as much as distinguish them. I am, in other words, intentionally reading Eliot backward: her philosophy of poetic form gives us a language for thinking about rocks as forms.

None of these theorists of form—Eliot, Levine, Macpherson, Rorty, Michaels—are explicitly interested in scale as such, but their convergence around questions of materiality, arrangement, and rocks gives us useful ways of thinking about geological scale as a legible form. Geology's formalism subsists not only in its deployment of uniformitarian or catastrophist narrative genres, but also in its Rorty-like reading of autopoietic rocks or, as geologists put it, "formations." Attuned to these conceptions of form as a relationship between materiality and pattern, we might notice that one of the most interesting formal features of the GTS, from the nineteenth century to the present day, is that its units and boundaries are simultaneously material and abstract, a duality that resonates with a familiar oscillation between form as the shape of matter (Aristotle) and form as an idea or arrangement (Plato). The reasons for this duality are partly historical: the original GTS was a scale without numerical units; strata could be correlated with periods, but the exact number of years reflected was a matter of significant controversy.[17] Hence, the GTS is defined in two ways: in terms of "geochronic units," or spans of time defined in millions or billions of years before 1950, and in terms of "chronostratigraphic units," the geological strata that correspond to geochronic units.[18] Zalasiewicz, who has argued for unifying these two scales, observes that "conceptually, there has been a 'dual and parallel hierarchy' of chronostratigraphic (time-rock) units used to designate rock bodies that formed contemporaneously and geochronologic (or time) units used to designate intervals in which they formed or during which other events occurred (e.g., evolution, extinction, deformation, transgression)."[19] It is for this reason that it is possible to mark a point in an abstract scale of time materially with a golden spike.

It is not only the case that literary formalisms offer a starting point for conceptualizing the materiality of the GTS. The legible strata of the Earth may also torque literary and philosophical conceptions of form by dramatizing the capacity of nonhuman agencies to construct meaningful scales. A stratum represents an instance, to borrow Macpherson's language, of matter taking shape. To a certain extent, this geological autopoiesis undoes Eliot's distinction between intrinsic and extrinsic forces as constituting different kinds of forms. While it is surely the case that the forces that create sediment layers are, strictly speaking, extrinsic and inorganic, they have the effect of producing rock formations that are internally coherent and legible. Strata, in their relations, become like the "higher" forms of a poem or a person in that if you divided them up and scattered them, they would no longer mean the same thing. This nonhuman production of meaning is unremarkable to a geologist; one might even say that if humanists have only begun to robustly theorize their posthumanisms, geologists have long occupied a parallel "prehumanist" territory in which the Earth is forced to yield meaning.

But the Anthropocene makes a difference for geoformalism. Until recently, geological formalism could be straightforwardly understood as prepolitical, because the scales it interpreted preceded human life. The notion of the Anthropocene as a unit of the GTS complicates the interface of form and geology. This is not only because discussions of the Anthropocene lead scientists to construct narratives about the intersection of human and Earth history, thereby reanimating literary forms that were mostly abandoned when geology turned to a classificatory and quantitative project in the mid-nineteenth century. It is also because the Anthropocene may unsettle the distinction between the human and nonhuman formation of strata, thereby introducing into geological formations a mode of legibility that mixes human and nonhuman inscription.

To understand how the Anthropocene complicates the relationship between geological and literary form, it is useful to attend to how certain geologists have contested the concept of the Anthropocene on the basis of its present illegibility in the rock record. The geologist Stanley C. Finney, a prominent member of the International Commission on Stratigraphy, has voiced trenchant objections to the Anthropocene: it is a stratigraphic era without content, since human deposits are at most "a few tens of centimeters thick"; it misunderstands the stratigraphic record, which "is studied in order to *interpret* past events"; and it requires speculative "projections into future millennia."[20] All of this leads Finney to propose that the Anthropocene is "of similar character to the term Renaissance": meaningful as a human-historical category but useless to geologists ("The 'Anthropocene' Epoch," 8). Other geologists share his skepticism; Whitney Autin and John Holbrook note that the Anthropocene inverts stratigraphic practice: "we are left to map a unit conceptually rather than conceptualizing a mappable stratigraphic unit."[21] Clive Hamilton, by contrast, views these kinds of objections, which "fetishize" a physical golden spike, as evidence that the paradigm of stratigraphy is "unsuited to making a judgement about the Anthropocene."[22] What these debates make apparent is an unresolved question about the material presence of the geological scale, including the ways in which human activity becomes legible within it. The Anthropocene represents an instance in which meaning precedes its inevitable material registration, and so stratigraphers are left with a situation in which their interpretation is determined in advance even as the text remains unwritten. Stratigraphy's observational practice uncomfortably edges closer to what Mitchell describes as formalism's politicized "reflection on the precise way things are said or shown": geological formations now speak to and display a story that involves human questions of responsibility, values, and politics.

Before taking up the question of how this understanding of scale as a substantial form might guide practices of reading novels as well as strata, it is worth observing that the intersection of literary and geological formalisms reveals one boundary of Levine's notion of form, which otherwise looks so capacious as to be dangerously portable: nearly everything, for Levine, can be called a form. But perhaps because she takes inspiration from design theory (form is defined by its "affordances") and seeks to find formalism's purchase on political life, Levine's conception of form is unapologetically anthropocentric; her instances of forms are more like Eliot's poems and persons—interconnected, intrinsically determined—than like rocks. It might initially seem that geologists' understandings of scale as a form, first materialized in the Earth and then interpreted via charts, narratives, and maps, would lead, by contrast, toward a nonhuman formalism. But this neat division of anthropogenic and non-anthropogenic forms is just what is complicated by the advent of the Anthropocene, which introduces the thought of geological strata that will be mixtures of human and nonhuman depositions. Geological scale thus becomes imbricated with human meaning and value, giving rise to a now-familiar conundrum: placing the rung of the Anthropocene on the ladder of eras and epochs is an interpretive act that attributes responsibility and generates narratives with political force. Scales imply values.[23] For Levine, the politics of form have to do with the ways that "the multiple forms of the world come into conflict and disorganize experience" (*Forms*, 17); Levine's "collisions" (18) most often involve sociopolitical and aesthetic formations (in a scene in *Jane Eyre*, for instance, the "hierarchy of the school" comes into conflict with the "rules of the courtroom") (21). Humans' arrival on the scale of geological time is a collision of forms not anticipated by Levine's categories: a collision that takes place not between human social and aesthetic forms (e.g., law, education, the novel) but between human and nonhuman scales.

### Scalar Failure

What would it look like to adopt a formalist interpretive practice oriented by this suggestive interface of questions about form in geology and in literary studies? It would almost certainly involve positioning concepts of scale at the center of formalism. One might begin to think of literary forms as scalar "actors," in that they teach readers to inhabit simultaneous but incommensurate scales of time and space.[24] In fact, formalist criticism has often theorized the scalar multiplicity of narrative: story time and narrative time almost always unfold at divergent scales—hence Gérard Genette's exploration of "the various types of discordance between . . . story and narrative" or Christian Metz's observation that "one of the functions of narrative is to invent one time scheme in terms of another time scheme."[25] Disjunctures and dilations of narrative time scales—"anachronies" and "anisochronies" in Genette's language—are rarely disorienting; to the contrary, this scalar multiplicity is among the formal givens of narrative, a readily identifiable structure that is common to diverse narrative media and transhistorically persistent.[26] Alongside reflections about how interpretation itself moves within various historical scales, the multidurational and anachronistic qualities of narrative and aesthetic forms have led scholars to explore how the Anthropocene might invite different ways of scaling literary and cultural interpretation, either by drawing attention to "scale variance" and unscalability, by "reading at several scales at once," by "thinking about scale historically as a language that determines how we formulate our origins," or by opening a space for "epochal, geologic time" in twenty-first-century fiction.[27]

It is often the case that arguments along these lines, which seek to secure a place for literary criticism in conversations about the Anthropocene, focus attention on how literary form can successfully respond to the difficulty of thinking at vast and multiple scales. It is in this spirit that Timothy Morton argues that certain artworks may "evoke" or "capture" the massive, imperceptible, and dispersed qualities of climate change.[28] But, turning now to the novels of Thomas Hardy, I explore the possibility that the opposite may be true. Although almost all of Hardy's novels address the interactions between human and nonhuman time scales, their ultimate significance may be that they symptomatically register failures to reconcile these two scales of time. I broadly agree with those who have argued that a new scalar criticism is necessary in the context of anthropogenic climate change. What I wish to show through my reading of Hardy is that a scalar formalism would attend to imaginative failures to bridge what Chakrabarty calls cognitive "rifts" ("Climate and Capital," 3)—rather than freight literary form with the task of bonding them.

Hardy's own reflections on his novels indicate that they are fertile territory for a formalist criticism interested in the problem of reconciling multiple scales, in that Hardy draws explicit connections between the conventions of Victorian narrative genres and scientific expansions of time and space. An 1888 notebook entry envisions a new kind of sensation novel whose excitement would result not from a rapid succession of "personal adventure" (as was true of bestsellers such as Wilkie Collins's *The Woman in White* [1860] and Ellen Wood's *East Lynne* [1861]) but rather from a "psychical" effect on the faculties: "sensationalism is not casualty, but evolution; not physical but psychical."[29] Hardy often realized this sensationalism of psychology rather than incident by placing the vast scales of nonhuman time and space in direct conflict with the comparatively small scale of human desire. In a letter to the critic Edmund Gosse, Hardy writes that his aim in the novel *Two on a Tower* was "to make science, not the mere padding of a romance, but the actual vehicle of romance": "vehicle" here implies a container, a means, or a medium; it is a word, the *Oxford English Dictionary* reminds us, that may mean the "form, the material or other shape, in which something spiritual is embodied or manifested."[30] The novel's preface elucidates what it would mean for science to become a vehicle; astronomy affords a comparison of the size of a dyad of lovers with the size of the universe: "This slightly-built romance was the outcome of a wish to set the emotional history of two infinitesimal lives against the stupendous background of the stellar universe, and to impart to readers the sentiment that of these contrasting magnitudes the smaller might be the greater to them as men."[31] Evoking these two vastnesses—the intensive vastness of psychic life and the extensive vastness of the heavens—the durational techniques of literary form became Hardy's means for dramatizing the incommensurabilities between scales of human value and of nonhuman interstellar space.

Hardy's comments about science as a "vehicle" suggest that many of the features of his novels that a reader might be apt to understand as thematizing scientific ideas instead reflect an aspiration to integrate scientific thought with literary form. For example, in Hardy's *A Pair of Blue Eyes*, the time of the sensational plot is frequently contrasted with the characters' rocky surroundings, which are finely observed with a geologist's eye. At one moment, Hardy recognizes the "vertical cuttings in metamorphic rock" for a railway; elsewhere he notices the quartz stone upon which one character's foot rests as "an igneous protrusion into the enormous masses of black strata, which had since been worn away from the sides of the alien fragment by centuries of frost and rain." England's southwestern cliffs are "stacks of rock."[32] A hill is not a hill; it is "a vast stratification of blackish-grey slate" (*Pair of Blue Eyes*, 192). One character loses a jewel in the "stone and crannies" of a "windy range of rocks" (62); later, it is found in the landscape's "innermost rifts and slits" (283). J. A. Pasquier's illustrations in *Tinsley's Magazine*, where the novel was serialized, reminded readers of the rocky setting, often portraying melodramatic moments when the love interest, Elfride, stands on the stony precipices of cliffs and buildings. This attention to rock brings into contact the two aspects of form I described above: its durational multiplicity (i.e., anachronies and rhythms) and its hylomorphic inseparability from shaped matter.

The rocky landscape serves as a material formation against which the temporal structure of the sensation narrative is brought into relief: the slow time of stone is a surface on which the characteristically rapid sensation plot is inscribed. A local precipice (dramatically called the "Cliff Without a Name") most clearly conforms the shape of narrative time to the shape of stone (*Pair of Blue Eyes*, 195). It is often suggested that narrative theory's most common geological metaphor, the "cliffhanger," originated with an installment-ending scene from Hardy's work when a character, Knight, slips over the edge of the cliff. This claim is likely apocryphal, but it importantly recognizes the congruity of geological and narrative form for Hardy. The "igneous protrusion" of quartz, whose surrounding shale has worn away over centuries, suddenly comes loose under the pressure of Knight's foot as he hoists Elfride to safety. Stone here becomes a site of contact between two widely disseminated Victorian narrative scales: the unimaginably slow geological change over time as propounded by Lyell's and Darwin's uniformitarianism; and the implausibly rapid incidents and coincidences plotted by sensation novelists, such as Wilkie Collins and Mary Elizabeth Braddon.

This contrast between the tempos of human and geological drama is rearticulated as a problem of human value and meaning in the scene that follows: Knight, dangling from a cliff, finds himself looking at a fossilized trilobite that happens to be at eye level. As Knight gazes at the fossil, he reflects on the differing scales of geological time and human time: "Separated by millions of years in their lives, Knight and this underling seemed to have met in their place of death." Knight looks at his stony surroundings and observes, "The immense lapses of time each formation represented had known nothing of the dignity of man" (*Pair of Blue Eyes*, 200). What these formations convey is not, as for Darwin, a sense of the vastness of geological time, but rather its deprecated relation to human life, and it is here that geological scale poses a problem of value as well as a problem of time: "They were grand times, but they were mean times too, and mean were their relics" (200). This is an echo of Charles Dickens's *A Tale of Two Cities* perhaps, phrased in terms of Earth history rather than human history. Time without humans is grand in its vast sweep, but because it is time without humans it is also time without "dignity"; it is "mean." Hardy's aim is not just to emphasize the vastness of geological time as inscribed in "formations"; it is to understand the advent of human systems of value in a geologic time scale. And yet there is a suggestion that some version of value did exist before humans, which takes the form of the trilobite's implied capacity to care about its existence. The trilobite is described as the only visible thing "that had ever been alive and had had a body to save, as he himself had now" (200). Knight's meditation on the relationship between geological scale and value involves antinomies: dignity exists and does not exist; time without humans is at once mean and grand. If we take seriously Hardy's aspiration to merge scientific thought and literary form, then we see that this scene does not just "represent" or "mediate" scientific ideas; rather, its radically extended anachrony, which situates millions of years of Earth history as the backstory of Knight's confrontation with death, uses the material duration of stone to deform the eventful rhythm of the popular fictional genres that Hardy would later describe as a sensationalism of "personal adventure."[33]

Hardy's interest in how form mediates the value effects of intersecting human and nonhuman scales is dramatized even more clearly in *Two on a Tower*, which involves a scalar incommensurability—that of astronomy—that self-consciously strains the limits of narrative form to contain vast scales. As in *A Pair of Blue Eyes*, the differing scales of human and nonhuman time produce a structuring narrative tension. In *Two on a Tower*, the melodrama of the amateur astronomer Swithin St. Cleeve's love affair with his patron, Lady Viviette Constantine, unfolds against the background of Swithin's painstakingly gradual observations of the night sky. This animates the familiar astrological literary imaginary of star-crossed lovers—but reverses it by portraying the sky as incapable of offering portents or human meaning. To Viviette's speculation that there are probably thousands or hundreds of thousands of stars, Swithin explains that through a telescope one can see "[t]wenty millions. So that, whatever the stars were made for, they were not made to please our eyes. It is just the same in everything: nothing is made for man" (31). Mimicking Knight's visual encounter with the millions of years prior to human life, Swithin's telescope brings into view a universe to which human life is peripheral. "Dignity" is destroyed by vastness, Swithin says: "There is a size at which dignity begins, ... further on there is a size at which grandeur begins, further on there is a size at which solemnity begins, further on a size at which awfulness begins, further on a size at which ghastliness begins. That size faintly approaches the size of the stellar universe. So am I not right in saying that those minds who exert their imaginative powers to bury themselves in the depths of that universe merely strain their faculties to gain a new horror?" (34). Hardy measures the scale of interstellar space with aesthetic categories—grandeur, awfulness, ghastliness—rather than numbers, hinting at an incommensurability between what is quantified and what is felt. What is most horrifying about the night sky is not what you can see, but what you cannot: "there are things much more terrible than monsters of shape; namely, monsters of magnitude without known shape"; moments later, Swithin reiterates the immense "size and formlessness" of the heavens (33, 34). Hence, it is not only that stellar magnitude exceeds numerical concepts; the felt extremities of vastness can be figured only in terms of shapelessness or formlessness, the outer limits of figuration itself.

It might seem, then, that Hardy is a preeminent theorist of how literary forms adapt themselves to scalar incommensurabilities. But as his scale of aesthetic categories begins to suggest, his novels ultimately cannot help but give form to vast formlessness. A familiar story about Hardy's scalar imagination is that it attempts, in Gillian Beer's words, to "find a place for the human" within the new scales of time opened up by evolutionary thought.[34] This is the interpretation for which Hardy prepares readers when he claims that *Two on a Tower* reveals the apparently "smaller" magnitude of human drama to be "greater" than interstellar magnitude. It is often observed that Hardy with virtuosity effects scalar shifts in visual perspective. Reflecting on Hardy's poem *The Dynasts*, John Plotz describes Hardy's "ability to leap upward from the microscopic to the sidereal and plummet earthward again with equal ease."[35] Anna Henchman similarly proposes that Hardy's "narrative experiments illustrate the ways in which the imagination can move quickly between vastly different standpoints, inducing a kind of productive perplexity rather than epistemological breakdown."[36] While these readings deftly elucidate Hardy's technical facility at shifting perspective, they may also be too willing to take the novels at their word. What if, instead, the narrative adeptness of these scalar shifts masks a conceptual failure? What if the novels appear to address nonhuman scales, but in fact fail to conceptualize them as nonhuman? It might then be more accurate to interpret Hardy's shifts of scale as seductive but insufficient attempts to grasp magnitudes of time and space that "annihilate" human life (as Viviette puts it): insufficient precisely because they rephrase these vast scales in terms of human judgments of "ghastliness" or "awfulness," shape them into a novel, and thus return them to comprehensibility. The "ease" or "productive perplexity" that Hardy's readers identify would then not be a solution, but a problem: narrative strategies of perspectival shifting and temporal rescaling too readily compensate for the real impossibility of non-anthropocentrically dwelling with scales of deep time or deep space. In the hands of the novelist, formlessness itself becomes a form.

This reading of Hardy as failing to encounter vastness through the very act of attempting to represent it is positioned at the intersection of symptomatic and formalist interpretation insofar as it identifies narrative features such as asynchronies and perspectival shifts as imaginary resolutions of philosophical contradictions. Hardy's attempt to integrate deep time and space with narrative structure is instructive in the context of the Anthropocene not because it is an aesthetic technique that supplements those of the geologists—novels as literary translations of the GTS—but rather because it calls attention, through its failure to escape anthropomorphisms, to the real epistemic challenges of nonhuman time scales. Hardy cannot but refract nonhuman scales through a language of human value, even in order to express them as lacking this value. This failure of Hardy's novels is not a failure on the part of Hardy; rather it reflects something deeper about the paradox of bringing nonhuman scales into human frames of reference. Eugene Thacker argues that thinking about human extinction may arrive at the limit of thought itself; glossing Immanuel Kant's *The End of All Things* (1794), Thacker proposes that "[i]f the idea of human extinction necessitates the negation of all thought, including the thought of this thought, then it would seem that extinction is not just a scientific concept, but a sort of limit-concept for all thought."[37] Kant's essay imagines this paradox in terms of a future extinction, but one could easily turn this insight around as speaking to why it is so difficult to conceive of an Earth before human life. It is a problem that geologists have long been familiar with; recall Darwin's language of cognitive failure: "the mind feebly . . . comprehend[s]" geological time; it forms only "an inadequate idea of the duration of each formation"; there are "an infinite number of generations, which the mind cannot grasp" (*Origin of Species*, 208, 210).

If it is true that Hardy fails exactly where many have seen him as succeeding, then this may open up a different way of thinking about what a renewed focus on aesthetic or literary form might bring to a multidisciplinary dialogue about scale and climate change. Chakrabarty argues that "rifts" or "gaps" appear between "the vastness of [the climate problem's] non- or inhuman scale . . . and how we think about it when we treat it as a problem to be handled by the human means at our disposal" ("Climate and Capital," 3). Critics and novelists including Timothy Morton, Ursula Heise, and Amitav Ghosh have proposed various ways in which literary and aesthetic forms may address this type of scalar disjuncture between immediate experience and scientific knowledge.[38] This is also the premise or subtext of many novels in the rapidly expanding field of climate fiction, which attempts to bring the scale of the crisis into the domain of the imaginable, for example, Kim Stanley Robinson's Science in the Capital trilogy (2004–7), Ben Lerner's *10:04* (2014), and Claire Vaye Watkins's *Gold Fame Citrus* (2015).

In theory and in practice, the literary imagination is now often called upon to visualize the climate crisis and its effects. Hardy reveals that, at least when it comes to reconciling multiple scales, this may be a task for which the novel is not well fitted. In this respect, Kant's discussion of the sublime may allow us to understand why aesthetic form would seem to be simultaneously exceptionally well suited to and also incapable of reconciling scalar incommensurabilities. Reflecting on how looking into telescopes and microscopes induces difficulties of judging size, Kant proposes that "the very inadequacy of our faculty for estimating the magnitude of the things of the sensible world awakens the feeling of a supersensible faculty in us"; this faculty, reason, is "absolutely great" and "surpasses every measure of the senses."[39] The experience of the sublime does not join sensibility and rationality; it is rather a "vibration" between the failure of the senses and the triumph of reason (*Critique*, 141). Notably, Kant resolves the problem of reconciling the magnitude of the stellar universe and the magnitude of numbers not by bridging scales of sense and reason, but by valuing reason over sense. This suggests that aesthetic form may only ever appear to bridge the rifts Chakrabarty describes by reducing vastness to human size.

This leaves us with a question about the relationship between scale and value that seems genuinely difficult. One reason that problems of scale have drawn attention in the context of the Anthropocene is that understanding anthropogenic climate change requires folding systems of value that are meaningful at the scale of human history into accounts of nonhuman Earth processes that are meaningful at the scale of Earth history. When Ian Baucom or Patricia Yaeger proposes that we must rethink literary history in relation to rising temperatures in the future or sources of energy in the past, they ask us to find a bridge between cultural objects with human meaning and features of the nonhuman world that constrain (or will constrain) many aspects of human agency, including the agency to create cultural forms.[40] Formalism, I have argued, may be an important tool in discovering these bridges, insofar as its recent varieties are attuned to the close relationship between aesthetic and material structure and afford ways of recognizing patterns and rhythms shared across seemingly divergent domains. But Hardy's novels, through their apparent brilliance at multiscalar thinking, alert us to the possibility that at the very moment when literary and aesthetic forms seem to be vividly bringing nonhuman scales into the realm of perception, they may instead be demonstrating our failure to apprehend them.

## Notes

1. Dipesh Chakrabarty, "Climate and Capital: On Conjoined Histories," *Critical Inquiry* 41.1 (2014): 1–23, 3 (hereafter cited parenthetically in text).

2. Mary C. Stiner et al., "Scale," in *Deep History: The Architecture of Past and Present*, ed. Andrew Shyrock and Daniel Lord Smail (Berkeley: University of California Press, 2012), 242–72; Julia Adeney Thomas, "History and Biology in the Anthropocene: Problems of Scale, Problems of Value," *American Historical Review* 119.5 (2014): 1587–607.

3. Martin Rudwick, *Bursting the Limits of Time: The Reconstruction of Geohistory in the Age of Revolution* (Chicago: University of Chicago Press, 2005), 181; Ralph O'Connor, *The Earth on Show: Fossils and the Poetics of Popular Science, 1802–1856* (Chicago: University of Chicago Press, 2007), 4. See also Martin Rudwick, *Worlds Before Adam: The Reconstruction of Geohistory in the Age of Reform* (Chicago: University of Chicago Press, 2008).

4. Adelene Buckland, *Novel Science: Fiction and the Invention of Nineteenth-Century Geology* (Chicago: University of Chicago Press, 2013), 2, 18. For a contrasting view, see James Secord: the "greatest accomplishment" of mid-nineteenth-century geology was no longer the construction of dramatic cosmological narratives of deep time; it was more simply "establishing the fundamental order of the world's rocks." Secord, *Controversy in Victorian Geology: The Cambrian-Silurian Dispute* (Princeton: Princeton University Press, 1986), 4.

5. John Phillips, *A Treatise on Geology* (London: Longman, 1837), 1:9 (hereafter cited parenthetically in text).

6. Arthur Holmes, *The Age of the Earth* (London: Harper, 1913), 10.

7. Charles Darwin, *On the Origin of Species*, ed. Gillian Beer (1859; repr., New York: Oxford University Press, 2008), 208 (hereafter cited parenthetically in text).

8. F. M. Gradstein and J. G. Ogg, "Time Scale," in *Encyclopedia of Geology*, ed. Richard C. Selley et al. (Oxford: Elsevier, 2005), 503.

9. Gary Nichols, *Sedimentology and Stratigraphy*, 2nd ed. (Chichester, England: Wiley-Blackwell, 2009), 297.

10. J. A. Zalasiewicz and Mark Williams, *The Goldilocks Planet: The Four Billion Year Story of Earth's Climate* (Oxford: Oxford University Press, 2012), xvii; Jan Zalasiewicz, *The Earth After Us: What Legacy Will Humans Leave in the Rocks?* (Oxford: Oxford University Press, 2008), 1.

11. Sandra Macpherson, "A Little Formalism," *English Literary History* 82.2 (2015): 385–405, 392. As Macpherson notes, rocks, stones, and trees mark the category of the unformed in Walter Benn Michaels, *The Shape of the Signifier: 1967 to the End of History* (Princeton: Princeton University, 2004), 82–128.

12. Macpherson, "A Little Formalism," 390.

13. Caroline Levine, *Forms: Whole, Rhythm, Hierarchy, Network* (Princeton: Princeton University Press, 2015), 3 (emphasis in original) (hereafter cited parenthetically in text).

14. W. J. T. Mitchell, "The Commitment to Form; or, Still Crazy After All These Years," *PMLA* 118.2 (2003): 321–25, 324.

15. Richard Rorty, "Philosophy Without Principles," *Critical Inquiry* 11.3 (1985): 459–65, 460.

16. George Eliot, "Notes on Form in Art," in *Essays of George Eliot*, ed. Thomas Pinney (New York: Columbia University Press, 1963), 434 (hereafter cited parenthetically in text).

17. Darwin argued that the geological record was fragmentary; John Phillips countered that this emphasis on its imperfection was "overrated." Phillips, *Life on the Earth: Its Origin and Succession* (London: Macmillan, 1860), 207.

18. The chart is discussed in the official publication F. M. Gradstein, *The Geologic Time Scale 2012* (Amsterdam: Elsevier, 2012).

19. Jan Zalasiewicz et al., "Chronostratigraphy and Geochronology: A Proposed Realignment," *GSA Today* 23.3 (2013): 4–8, 4.

20. Stanley C. Finney and Lucy E. Edwards, "The 'Anthropocene' Epoch: Scientific Decision or Political Statement?" *GSA Today* 26.3 (2016): 4–10, 7, 8 (hereafter cited parenthetically in text); Stanley C. Finney, "The 'Anthropocene' as a Ratified Unit in the ICS International Chronostratigraphic Chart: Fundamental Issues That Must Be Addressed by the Task Group," *Geological Society, London, Special Publications* 395.1 (2014): 23–28, 26.

21. Whitney J. Autin and John M. Holbrook, "Is the Anthropocene an Issue of Stratigraphy or Pop Culture?" *GSA Today* 22.7 (2012): 60–61, 61.

22. Clive Hamilton, "Getting the Anthropocene So Wrong," *Anthropocene Review* 2.2 (2015): 102–7, 106.

23. Deborah Coen outlines the value effects of scaling in the Anthropocene in "Big Is a Thing of the Past: Climate Change and Methodology in the History of Ideas," *Journal of the History of Ideas* 77.2 (2016): 305–21.

24. I borrow the idea of literary texts as "nonhuman actors" from Rita Felski, *The Limits of Critique* (Chicago: University of Chicago Press, 2015), 162–72.

25. Gérard Genette, *Narrative Discourse: An Essay in Method* (Ithaca: Cornell University Press, 1983), 36; Christian Metz, *Film Language: A Semiotics of the Cinema* (Chicago: University of Chicago Press, 1991), 18.

26. Genette calls attention to the fact that *The Odyssey* depends extensively on anachronies (*Narrative Discourse*, 47–49). On the notion of form as a "given," see Frances Ferguson, "Jane Austen, *Emma*, and the Impact of Form," *Modern Language Quarterly* 61.1 (2000): 157–80.

27. Derek Woods, "Scale Critique for the Anthropocene," *Minnesota Review* 2014.83 (2014): 133–42, 133; Timothy Clark, "Derangements of Scale," in *Telemorphosis: Theory in the Era of Climate Change*, ed. Tom Cohen (Ann Arbor: Open Humanities, 2012), 163; Noah Heringman, "Deep Time at the Dawn of the Anthropocene,"

*Representations* 129.1 (2015): 56–85, 61; Kate Marshall, "What Are the Novels of the Anthropocene? American Fiction in Geological Time," *American Literary History* 27.3 (2015): 523–38, 524. Though not explicitly connected to climate change, Wai Chee Dimock and Mark McGurl's discussions of deep time as an interpretive context is relevant. See McGurl, "The Posthuman Comedy," *Critical Inquiry* 38.3 (2012): 533–53; Dimock, "Low Epic," *Critical Inquiry* 39.3 (2013): 614–31.

28. Timothy Morton, *Hyperobjects: Philosophy and Ecology After the End of the World* (Minneapolis: University of Minnesota Press, 2013), 184, 185.

29. Florence Emily Hardy, *The Early Life of Thomas Hardy, 1840-1891* (New York: Macmillan, 1928), 268.

30. Thomas Hardy, *The Collected Letters of Thomas Hardy*, ed. Richard Little Purdy and Michael Millgate (Oxford: Clarendon, 1978), 1:110; *Oxford English Dictionary Online*, s.v. "vehicle."

31. Thomas Hardy, *Two on a Tower*, ed. Suleiman M. Ahmad (1882; repr., Oxford: Oxford University Press, 1998), 3 (hereafter cited parenthetically in text).

32. Thomas Hardy, *A Pair of Blue Eyes* (1873; repr., Oxford: Oxford University Press, 2009), 194, 299 (hereafter cited parenthetically in text).

33. Hardy, *Early Life of Thomas Hardy*, 268.

34. Gillian Beer, *Darwin's Plots: Evolutionary Narrative in Darwin, George Eliot, and Nineteenth-Century Fiction*, 3rd ed. (Cambridge: Cambridge University Press, 2009), 235.

35. John Plotz, "Speculative Naturalism and the Problem of Scale: Richard Jefferies's *After London*, After Darwin," *Modern Language Quarterly* 76.1 (2015): 31–56. On Hardy, Darwin, and scale, see also Beer, *Darwin's Plots*, 220–41.

36. Anna Henchman, *The Starry Sky Within: Astronomy and the Reach of the Mind in Victorian Literature* (Oxford: Oxford University Press, 2014), 157.

37. Eugene Thacker, "Notes on Extinction and Existence," *Configurations* 20.1 (2012): 137–48, 144–45.

38. Morton, *Hyperobjects*, 159–201; Ursula K. Heise, "Lost Dogs, Last Birds, and Listed Species: Cultures of Extinction," *Configurations* 18.1 (2010): 49–72; Amitav Ghosh, *The Great Derangement: Climate Change and the Unthinkable* (Chicago: University of Chicago Press, 2016).

39. Immanuel Kant, *Critique of the Power of Judgment*, trans. Paul Guyer (Cambridge: Cambridge University Press, 2000), 134 (hereafter cited parenthetically in text).

40. Ian Baucom, "History 4°: Postcolonial Method and Anthropocene Time," *Cambridge Journal of Postcolonial Literary Inquiry* 1.1 (2014): 123–42; Patricia Yaeger, "Editor's Column: Literature in the Ages of Wood, Tallow, Coal, Whale Oil, Gasoline, Atomic Power, and Other Energy Sources," *PMLA* 126.2 (2011): 305–26, 305–10.

# 8.
# Anthropocene Interruptions
## Energy Recognition Scenes and the Global Cooling Myth

*Justin Neuman*

One of the more insidious arguments used by skeptics and deniers of global warming to challenge the scientific consensus about the rate and direction of climate change has to do with a theory, advanced in the 1970s, that anthropogenic pollutants in the form of aerosols and particulates could increase atmospheric reflectivity, resulting in potentially catastrophic global cooling. This hypothesis, really a countertheory advanced against the growing consensus around the role of carbon dioxide and methane emissions in "greenhouse" effects, gained traction in the popular scientific press after the publication of an article in *Science* in July 1971 arguing that "an increase by only a factor of 4 in global aerosol background concentration may be sufficient to reduce the surface temperature by as much as 3.5 degrees Kelvin . . . sufficient to trigger an ice age."[1] The story resonated well with a U.S. audience already primed by a series of exceptionally cold winters in the 1950s and 1960s and whose fear of the predicted cold was nourished on a Cold War imaginary replete with theories and fictions of nuclear winter[2]—the hypothesis that the firestorms, dust, and fallout attendant to a large-scale nuclear conflict could block out the Sun and cool the Earth. Dozens of headlines in the early 1970s blared warnings, including "New Ice Age Coming" (*L.A. Times*, October 24, 1971). *Newsweek* published "The Cooling World" (April 28, 1975), prophesying a "drastic decline in food production"; and *Time*'s "Another Ice Age?" (June 24, 1974) hypothesized about the effects of climate change just as the Arab oil crisis amplified fears of cold winters and unaffordable heating oil.[3]

It turns out that the idea of a consensus around cooling trends in the 1970s is a myth, though at the time of this writing, headlines from that decade's global cooling scare had been resuscitated by right-of-center pundits, conservative talk-show personalities, and Republican presidential candidates, who pounced with glee on the allegation that many of the publications and organizations now calling for immediate and decisive action to combat global warming were, more than forty years ago, equally convinced that another ice age was a clear and present danger.[4] For those seeking action on climate change, theories of anthropogenic cooling in the 1970s are an inconvenient truth, and numerous websites and counterstudies debunk the supposed consensus on cooling and suggest talking points for activists hoping to distance themselves as quickly as possible from an embarrassing stumbling block.[5] In this chapter I follow an alternate trajectory, suggesting that theorists of the Anthropocene have much to learn from the earlier global cooling speculation and from the modes of energy systems analysis developed in the late nineteenth and early twentieth centuries, specifically that of American historian and diplomat Henry Adams, who predicted in 1910 the imminent and irreversible cooling of the planet. Against the presentism that characterizes much of the debate surrounding the Anthropocene, my contribution to this volume turns our gaze to the time before the Great Acceleration of the postwar period, identifying both the sources of our current crisis and strategies for securing a more sustainable future in the late Victorian and modernist cultural imaginary.

Henry Adams, a man who before he was six "had seen four impossibilities made actual,—the ocean steamer, the railway, the electric telegraph, and the Daguerreotype," experienced a revelation at the Exposition Universelle held in Paris during the summer of 1900.[6] Though he considered himself very much a product of the machine age and was no stranger to such spectacles, having "lingered long among the dynamos" at the World's Columbian Exhibition in Chicago seven years earlier, Adams "haunted" the Palais de l'Électricité in Paris, enthralled by the generators that transformed the "heat latent in a few tons of poor coal" into electrical energy through an "occult mechanism" (*Education*, 353). "To Adams the dynamo became a symbol of infinity," he writes in the distinctive third-person voice of *The Education of Henry Adams*, published privately in 1907, confessing that "he began to feel the forty-foot dynamos as a moral force, much as the early Christians felt the Cross. . . . Before the end [of the fair], one began to pray to it" (352). Later that winter, having recovered somewhat from the shock of the new, Adams identifies what he takes to be his own duty in response to the vast changes he is witnessing: "The historian's business was to follow the track of the energy; to find where it came from and where it went to; its complex source and shifting channels; its values, equivalents, conversions" (361). This moment is a prototype of an event I call an "energy recognition scene," a moment, often surprising or unexpected, when writers and artists imagine their way across the vivid and tangible materiality of locomotives, lightbulbs, automobiles, and other technologies to the energy systems on which they depend. These moments stage an emergent awareness of the various ways the experience of technology depends on supply chains and entails externalities that extend spatially and temporally beyond a text's representational systems. Such moments, I argue, interrupt the intrinsic experience of commodities, technologies, and built environments, opening texts to a new range of structural, emotional, and semantic intensities.

The idea that power and speed are central, indeed constitutive, preoccupations of modernist art—the story of how fossil fuels and electric technologies transformed the social, behavioral, and economic practices of industrial societies—was central to modernist self-understanding and has informed the field of modernist studies since its inception. But while energy-intensive technologies became ubiquitous in modern life and literature, writers have rarely seen it as their "business," despite Adams's entreaties, "to follow the track of the energy," and few scholars have seriously considered whether fiction has much to say on the subject. Instead, "pouring in petrol," as E. M. Forster puts it in the opening vignette of *Howards End*, remains for most writers one of many "actions with which [their] story has no concern."[7] But as Forster underscores with his use of paralepsis, the rhetorical strategy of passing over a point in order to emphasize it, it is precisely by concerning ourselves with the story of energy—not only with its sources, but also with the technical, social, and economic systems that naturalize patterns of consumption and make it possible to take energy for granted—that we can achieve the novel's epigrammatic mandate, "only connect." Heeding Adams despite his false cooling prophecy and responding to more recent calls to get "serious," as Patricia Yaeger admonishes in a 2011 issue of *PMLA*, in our attempts to theorize "about literature through the lens of energy,"[8] in this chapter I develop an energy systems approach to literature that entails the twofold task of (a) reading out from the traces of energy use encoded within a text to the externalities and converter chains that exceed those traces in space and time, and (b) reading in to analyze the discourses and images that condition subjective experiences of energy use and also contribute to its invisibility.

By the first decade of the twentieth century, it was obvious to Henry Adams that human agricultural and industrial activities—in particular the rising rates of fossil fuel use—had altered our planet on climatologically and geologically significant scales. In 1910, Adams called society to account for its role in causing what we would now call anthropogenic climate change in *A Letter to American Teachers*, where he focused specifically on the profligate use of natural resources:

> Man dissipates every year all the heat stored in a thousand million tons of coal which nature herself cannot now replace, and he does this only in order to convert some ten or fifteen per cent. [sic] of it into mechanical energy immediately wasted on his transient and commonly purposeless objects. He draws great reservoirs of coal-oil and gas out of the earth, which he consumes like coal. He is digging out even the peat-bogs in order to consume them as heat. He has largely deforested the planet, and hastened its desiccation. He seizes all the zinc and whatever other minerals he can burn, or which he can convert into other forms of energy, and dissipate into space . . . on a scale that rivals operations of nature.[9]

Against the pervasive optimism of his time, Adams seems to have understood viscerally and intellectually that the indiscriminate use of hydrocarbon resources would radically accelerate the collapse of both ecosystems and civilizations.

While Adams offers a prescient and in many ways prophetic critique of unsustainable resource use, a subject to which I return below, he draws a series of mistaken conclusions on the question of climate change. Citing the most up-to-date sources, including his own translations from French sources when necessary, Adams argues that "the glacial period is far from being ended. . . . this retreat of the ice [the current interglacial period during which all of recorded human history has unfolded] is not definitive, but . . . the cold will return, and with the depopulation of a part of our globe . . . during this cataclysm revolutions will occur which the most fecund imagination cannot conceive" (*Letter*, 72). For Adams, the climatological effect of fossil fuel consumption will result not from carbon dioxide emissions and radiative forcing but as a consequence of accelerating the natural entropic processes by which all energy degrades. "All nature's energies were slowly converting themselves into heat and vanishing into space," writes Adams, "until, at last, nothing would be left except a dead ocean of energy at its lowest possible level"; burning fossil fuels, he speculates, will hasten this ultimate end (10).

Seeking to understand the effects of fossil fuel use on the Earth system, Adams turns to physics, in particular to the second law of thermodynamics: the energy in a closed system always and inevitably dissipates; entropy alone increases.[10] The idea of entropy hits like a wrecking ball the tranquil spheres of "the Newtonian universe, in which they had been cradled, [which] admitted no loss of energy in the solar system" (*Letter*, 5). Against the idea of human progress bequeathed to him from the Enlightenment, "only the ash-heap was constantly increasing in size" (5). Quoting from William Thomson's (Lord Kelvin) "On a Universal Tendency in Nature to the Dissipation of Mechanical Energy" (October 1852), Adams extrapolates to the heat death of the solar system: "within a finite period of time past, the earth must have been, and within a finite period of time to come, the earth must again be, unfit for the habitation of man as at present constituted" (4).

Today's more accurate paleoclimatic data reveals that for much of the past several hundred thousand years, as Adams suspected, ours has been an undeniably frosty planet: the line of temperature change relative to the twentieth-century average (an important figure used by today's climatologists as a baseline against which to measure climate change) shows deep, broad valleys of cold (reaching minus ten degrees Celsius at points) with a small warm island rising up 130,000 years ago, a brief respite before the end of the last ice age 15,000 years ago.[11] Adams is technically right about the ultimate heat death of the solar system—a star's life is finite—but like many scientifically inclined intellectuals trying to understand the implications of the second law of thermodynamics in the first half-century after its formulation,[12] Adams radically underestimates the relevant time scales, chilling his audience with visions of humanity "driven southward before the ice-cap which obliterated every trace of him and of his polar Eden as he slowly drifted towards the fortieth parallel" (*Letter*, 59). In humanity's imminent future, Adams suggests, "equatorial humanity will undertake vain arctic expeditions to rediscover under the ice the sites of Paris, of Bordeaux, of Lyons [sic], of Marseilles" (74)—expeditions whose implied rigors and dangers would remind his audiences of the great age of polar exploration, then at its apogee and a frequent and popular exhibit subject at the world's fairs of the period. In other words, warns Adams, "winter is coming." This reference to the chilling line from George R. R. Martin's *A Game of Thrones* is made only half in jest. One need not appeal to complicated theories of evolutionary biology to observe that most people like being warm and dislike the cold. The fear that our world may end not in fire but in ice leads us not only to negative somatic associations but also toward a deep vein of entropic modernism, a mode that reaches its apotheosis in the early poetry of T. S. Eliot and creeps to a halt one degree above absolute narrative zero in the plays of Samuel Beckett.

Henry Adams was far from the only critic of industrialization, and modernist art is as much a rejection as an exploration of technological change, but among his Victorian contemporaries, critics of industrialization tended to focus on the impacts of a coal-based economy in terms of aesthetics, human health, and local environmental degradation. John Ruskin, for instance, rails against the railroad as a blight on the landscape and a threat to aesthetic experience, suggesting in his essay "The Moral of Landscape" that twelve miles covered on foot should be the maximum distance for a day's pleasure travel.[13] From the hellish conditions of the coal mines depicted in Émile Zola's novel *Germinal* (1885) to the critique of working-class living conditions in liberal British parliamentarian Charles Masterman's *The Condition of England* (1909), critics inveighed against the social and health effects of coal culture. Writing about the "Black District" from which Henry Adams fled in horror ("he ran away from everything he disliked"; *Education*, 72) when he toured the industrial zones of England in 1858, Masterman laments that a half century later, devastation remained the norm: "so is being heaped up the wealth of the world. Under darkened skies, and in an existence starved of beauty, these communities of men and women and children continue their unchanging toil. Is the price being paid too great for the result attained?"[14] "This is the Age of Energy," writes Frederick Soddy; more accurately, Soddy adds, "this is the beginning of the ages of energy, the Age of the Energy of Coal. . . . That this still is the age of the energy of coal is unfortunately only too true, and the whole earth is rendered the filthier thereby."[15]

Historian Lewis Mumford, whose 1934 volume *Technics and Civilization* remains a seminal analysis of the technological basis of modern society, argues that in the nineteenth century, "coal and iron were the pivots upon which the other functions of society revolved."[16] For Mumford, coal epitomizes the destructiveness of the stage of technological development he describes as "paleotechnic" in contrast to the "neotechnic" age of electricity: "in all its broader aspects, paleotechnic industry rests on the mine. . . . [F]rom the mine came the steam pump and presently the steam engine: ultimately the steam locomotive and so, by derivation, the steamboat. From the mine came the escalator, the elevator . . . the subway for urban transportation. The railroad likewise came directly from the mine" (*Technics*, 158). Though acknowledging the massive growth in population, trade, and material goods undergirded by the energy of coal, in Mumford's analysis, "the state of paleotechnic society may be described, ideally, as one of wardom. Its typical organs, from mine to factory, from blast-furnace to slum, from slum to battlefield, were at the service of death" (195). Soddy comes at the question from a different, less anthropocentric angle, noting that "not a thought is given as to what the coal has cost Nature.... it is not possible to reckon what the cost of coal has been in the economy of Nature, nor how many ages of future time will be necessary to recuperate the amount now burned in a single year" (*Matter and Energy*, 36–37).

At the turn of the twentieth century many scientists, social historians, and the popular media increasingly understood their era as defined not only by the energy intensity of its culture—a phenomenon whose roots reach back to the invention of the steam engine—but also by the increasingly rapid emergence of new energy regimes and transportation systems that, they believed, would displace King Coal. Moving forward from 1900, it took only three years for the late Victorians, who had only recently gotten the wheels turning on their cars, to take to the skies in controlled, powered flight. Two years later, in 1905, Albert Einstein established the theoretical equivalence between mass and energy, further dissolving the boundary between the science and fiction of power by priming public expectation for free and unlimited energy. Handbooks for aspiring automobilists from the first decade of the twentieth century already predicted the coming of the emission-free electric car. A. B. Filson Young writes in *The Complete Motorist*: "on the day when a cheap, light, and compact means of storing a great power of electricity is discovered we shall see the last of the motor-car as we know it at present."[17] Guided by their faith in progress—a faith well buttressed by the evidence of ever-increasing coal and steel production, the size and speed of trains and ships, and the range and power of their artillery (but blind to the negative externalities associated with these technologies)— even the most ecological and conservation-minded Victorians expected a rapid transition from dirty coal to a new and externality-free fuel that would provide humanity with free and unlimited energy. As Mumford and others understood it, electricity promised to revolutionize the sorry state of paleotechnic society, ushering in a neotechnic phase of cleaner, more sustainable technological growth and development. The fact that battery troubles still belabor the electric automobile industry and that the vast majority of humanity's energy needs are still met through the burning of fossil fuels means that the twenty-first century is still overwhelmingly mired in a paleotechnic energy system, with all of the consequences in terms of human and ecosystem health that system entails.

Inspired by a variety of theories about the energetic basis of human life, nineteenth-century writers embraced the idea of the electric text, drawn especially to the possible analogies between electrical energy and the figurative power of art, understood in terms of its capacity to "move" its audience. Electrical technologies inflect Ralph Waldo Emerson's theories of literary inspiration, specifically his concept of the poet as the "conductor of the whole river of electricity," and they proliferate in Walt Whitman's poetry of "the body electric" with its "instant conductors."[18] Likewise, in "The Painter of Modern Life," Charles Baudelaire imagines the flaneur entering the flow of city crowds "as though it were an immense reservoir of electrical energy."[19] Patrick Geddes, who coined the term "Second Industrial Revolution" to describe the structural changes ushered in by electrification, maintained a utopian fantasy in which hydropower would satisfy the energy needs of society, making new leaders of possible hydropower energy giants Norway, Sweden, and Switzerland.[20] Similarly, Frederick Soddy, a quixotic nuclear chemist and a founder of environmental economics who would be awarded the Nobel Prize in chemistry in 1921 for his work on radioactive isotopes, prophesied in 1912 that "it is possible to look forward to a time which may await the world, when this grimy age of fuel will seem as truly a beginning of the mastery of energy as the rude stone age of Paleolithic man now appears as the beginning of the mastery of matter" (*Matter and Energy*, 16).

Search as you might among the soaring concrete towers or along the elevated highways, rail links, and airports of Le Corbusier's unrealized 1924 design for a "radiant city," you will find no gas stations or power plants.[21] The fact that Le Corbusier takes energy for granted is not unique but rather the norm; humanity's energy demands are, more often than not, solved symbolically rather than practically in the modernist cultural imaginary, which is supported by a robust faith in technological progress and nature's boundlessness, convictions only recently beginning to fray in the face of the harsh realities of climatological tipping points, sea level rise projections, and carrying capacities. F. T. Marinetti's deliberately unsustainable attempt to recruit the biophysical properties of gasoline to power the aesthetic engine of futurism proves a limit case in this regard.[22] Of the energy regime powering Le Corbusier's radiant city, we can learn little more than we can of the futuristic mechanism that powers E. M. Forster's science fiction fable "The Machine Stops" (1909), in which we are told merely that man has "harnessed Leviathan," securing free and unlimited power.[23]

Films like Georges Méliès's transnational sensation *Le voyage à travers l'impossible* (1904) are emblematic of the period's mixed energy regimes and pervasive resource optimism: in this twenty-minute silent film depicting a geographical society's journey to the Sun and back using "all the known means of locomotion—railroads, automobiles, dirigible balloons, submarine boats, etc.," Méliès depicts not a single scene of fueling.[24] Clouds of steam, smoke, and flame are expelled from the radiators, smokestacks, and tail pipes of Méliès's impossible machines, but while such scenes of conspicuous resource consumption produce exhausts of waste, the film's inventors and voyagers never face problems of energy scarcity. Instead, Méliès's film is staged in a world of boundless—though potentially destructive—supplies of energy and is concerned primarily with the dangers of plenitude: fires too hot, engines too fast, and freezers too cold.

A pioneering example of narrative cinema and special effects, *Le voyage à travers l'impossible* highlights contrasts and continuities between the familiar technologies of the Industrial Revolution—foundries, steam engines, and railroads—and the utopian fantasies of rapid transportation promised in the century to come, including spacecraft and automobiles capable of moving at three hundred miles per hour. Méliès's film is thus a "projection" not only in terms of the moving picture technology it employs but also in a more figurative sense, as a vision or prophecy about seemingly imminent technological developments. Tableau 3, "The Machine Shop," provides a visual education in the nineteenth-century technoscape, framed in a setting that re-creates the iron-and-glass construction of the period's ubiquitous exhibition halls and comments on its penchant for displays of giant machinery. The action of the scene takes place in a setting meant to evoke the evolving phases of the energy system, with actors milling about between the modern dynamo tended by a mechanic in the foreground and the massive flywheel and piston of a steam engine—a mainstay of mid-nineteenth-century industrial mills—that comprises the set's immediate background. Also present though not visible in the frame are other recent inventions, including a submarine and a zeppelin; plainly visible through the glass windows of the machine shop, a painted vista of an iron truss bridge completes the visual paean to the achievements of modern technology. In the machine shop the engines are going "full blast. Everything is in motion; flywheels of steam engines, hammers, stampers, cranks, and pistons." The narrative's eccentric protagonist, a mad scientist, strides around his laboratory scribbling down theories and equations, pursued all the while by a servant who tries to get him to eat his breakfast. Like the machines in the background, however, the scientist needs no fuel: in a slapstick climax of exasperation he kicks the servant's tray of food into the air, a gesture of defiance, and returns to his calculations, sustained, like his machines, by creativity and insight alone.

In *Le voyage à travers l'impossible*, the trip to the Sun begins, ironically, from a faithful reproduction of a Paris metro station, complete with porters and a ticket counter, through which the members of the Institute of Incoherent Geography pass to board their "special train." While the film's whimsical, hand-painted frames emphasize comedy over credibility in their depiction of interplanetary space travel, Méliès is keen to stress the continuities between the extraordinary and the everyday, beginning with the travelers' route itself. Following the train as it leaves Paris, the film showcases the most technologically advanced rail projects of the era, traveling to Switzerland to survey the construction of the Jungfraubahn, a train route that eventually linked the Swiss village of Grindelwald with a mountaintop station, still Europe's highest at an elevation of 3,454 meters. Like Méliès's filmic journey, the purpose of the Jungfraubahn is tourism rather than transportation: construction of the railway began in 1896, with its long tunnel to the Eigerwand station opening to tourists on June 28, 1903, allowing Méliès's audience a view of the Alps from a dizzying perch high on the famed north wall of the Eiger, though the final section of track would not be completed until 1912. Rushing symbolically past the greatest technical achievements of the era, in the film the expedition's special train appears to leap off the summit of the Jungfrau, held aloft by two dirigible balloons while it soars past comets and constellations toward an anthropomorphized Sun.

The voyagers ultimately return to Earth unharmed, and the film ends with a grand procession celebrating their quixotic journey, offering Méliès another chance to catalog the technological achievements of the nineteenth century—from the piston engine to the dynamo, from the invention of the steam locomotive to a train to the top of Europe.

*Le voyage à travers l'impossible* naturalizes expectations for developments in transportation technologies, offering vivid testimony to the fact that for early motorists using internal combustion engines to power voyages across land, sea, and sky, the technical challenges of producing smaller, faster, cheaper, and more reliable boats, planes, and automobiles weighed far more heavily than the problem of securing fuel. Plentiful coal, cheap oil, and the possibility of nuclear power underwrote a sense of energy's easy accessibility in the early twentieth century. In modern industrial societies more broadly, "energy has been integrated into ... [our] thinking as primary data," the energy historians Jean-Claude

Debeir, Jean-Paul Deléage, and Daniel Hémery suggest. "It has been considered implicitly as neutral, unlimited, inexhaustible, like water and oxygen, and not only devoid of any particular impact on the future of society, but as subordinate to this future, adaptable at will."[25] To those at the end of the nineteenth century and the beginning of the twentieth, both the benefits and the costs of coal were all too clear, but whether viewed with revulsion by skeptics like Adams and Ruskin or as a testament to unbridled progress by utopians like Méliès and Soddy, the prospect of energy scarcity or the exhaustion of fossil fuel resources remained almost inconceivable. Only now are we coming to understand the implications of resource use beyond the mere fact of finitude. In light of anthropogenic climate change, a planet implicitly regarded as infinite in available resources and its ability to support our species appears suddenly fragile, finite, and vulnerable. Over the *longue durée*, human control of energy (understood as the capacity to do useful work) has largely been restricted to the amount of power that could be harnessed from available flows of sun, wind, and water and from the heat of burning biomass—a story circumscribed by the limits of muscle work provided by humans and other animals. The widespread use of fossil fuels—beginning with coal and subsequently oil and gas—transformed this relatively stable equation of energy inputs and production outputs, effectively transforming what the Abrahamic faiths take to be the defining feature of the postlapsarian condition: the unceasing demand for physical work.

Looking back with Henry Adams from the Paris exposition in 1900, ever-larger and more powerful ships, trains, and stationary motors testified to more than a hundred years of incremental development within the coal-steam-iron matrix. Though Adams got it massively wrong on the question of climate change, his writings on energy and the environment pushed back against what he saw as the unwarranted optimism of his age, railing against the consequences of the ever-increasing energy intensity of modern technoculture. Noting the frequent presence of steam power in the pastoral spaces of nineteenth-century American literature, Leo Marx argues in *The Machine in the Garden* that the locomotive had already become a "national obsession" by the 1830s: "It is the embodiment of the age, an instrument of power, speed, noise, fire, iron, smoke—at once a testament to the will of man rising over natural obstacles, and, yet, confined by its iron rails to a predetermined path, it suggests a new sort of fate."[26] In his quirky literary memoir *The Education of Henry Adams*, Adams claims to have "been born [in 1838] with the railway-system; had grown up with it; had been over pretty near every mile of it with curious eyes" (307).

Both the success and the longevity of the steam-coal system constitute a major focus for Adams, who was no neophyte to those temples of technology, the great exhibitions and world's fairs, when he "haunted" the hall of dynamos in Paris in the summer of 1900, "aching to absorb knowledge" (*Education*, 352). In fact, Adams spends more pages talking about the education he received at the World's Columbian Exhibition of 1893 (where he "found matter of study to fill a hundred years" and "lingered long among the dynamos" that powered the famous "City of Light") than he does on his schooling at Harvard (316). Adams was not a Luddite newly converted to the wonders of the industrial age, and yet he describes how in 1900 he found himself, figuratively speaking, "lying in the Gallery of Machines . . . with historical neck broken by the sudden irruption of force totally new" (355). In the famous analogy from which he takes the title of his chapter on the Paris fair, "The Dynamo and the Virgin," Adams is floored by the revelation that religion as a motive force for human action has been replaced by technology. What troubles Adams and drives him to fill, if we credit the claim, "thousands of pages" in his notebooks, is that while he understands how a steam engine can transform the thermal energy latent in coal into kinetic energy to power trains, ships, and factories, electrical energy, like radiation, proves a different matter entirely (362). "No more relation could he discover between the steam and the electric current," Adams writes, "than between the Cross and the cathedral" (352).

In biology, the social sciences, and environmental economics, energy systems analysis involves studying the total inputs, flows, and outputs of a system from the standpoint of energy and materials at scales ranging from the cellular level to the biosphere. An energy systems perspective gives Adams traction against the optimism of an age in which "society by common accord agreed in measuring its progress by the coal-output" (*Education*, 135). Adams remains unconvinced, arguing instead that "man is a bottomless sink of waste unparalleled in the cosmos, and can already see the end of the immense economies which his mother Nature stored for his support" (135). In Adams's view, though "any one might hire, in 1905, for a small sum of money, the use of 30,000 steam horse power to cross the ocean," these staggering forces are an illusion of power, concealing both social costs and "the energy of vegetable growth [that] reached its climax as early as the carboniferous [period]" (46). In this analysis, Adams anticipates the argument made by Will Steffen, Paul Crutzen, and John McNeill in one of the seminal articles of the contemporary climate change debate: the intensification of fossil fuel use constitutes "a massive energy subsidy from the deep past to modern society, upon which a great deal of our modern wealth depends."[27] Lewis Mumford puts it differently, connecting fossil fuels with imperialism: "in the economy of the earth, the large-scale opening up of coal seams meant that industry was beginning to live for the first time on an accumulation of potential energy, derived from the ferns of the carboniferous period, instead of upon current income. In the abstract, mankind entered into the possession of a capital inheritance more splendid than all the wealth of the Indies" (*Technics*, 157).

It is, I think, one of the great ironies of the modern environmental movement that widespread public acceptance of anthropogenic climate change has come at almost the precise moment when atmospheric carbon dioxide concentration has exceeded what many climatologists have identified as a tipping point in Earth's geophysical systems. The implications of this more than inconvenient truth is the subject of the message of the 2013 "Summary for Policymakers," the lead article of the *Fifth Assessment Report of the Intergovernmental Panel on Climate Change*, which closes with a sobering declaration on the subject of irreversibility: even in a scenario in which humanity could immediately and magically return greenhouse gas emissions to preindustrial levels, the effects of cumulative emissions since the beginning of industrialization mean that "most aspects of climate change will persist for many centuries even if emissions of $CO_2$ are stopped."[28] The resulting sense of belatedness gives rise to a cluster of psychological affects associated with climate change, including helplessness, dread, apathy, and resentment. Srinivas Aravamudan describes the temporal effects of climate change as a mode of "catachronism," a form of temporality that "re-characterizes the past and the present in terms of a future proclaimed as determinate but that is of course not yet fully realized"; we are, for Aravamudan, "after the clinamen."[29] In a similar vein, Bruno Latour speaks of how "we have also to absorb the disturbing fact that the drama has been completed and that the main revolutionary event is behind us."[30] In other words, with atmospheric carbon levels at more than 400 parts per million and rising, with the "carbon 350 target" advanced by Bill McKibben and others fading rapidly in the rearview mirror of our single-occupancy SUVs, we, insofar as such scenarios contain a "we," will be living on a warmed planet for quite some time.[31]

It seems unlikely that we can or will, as a species or as a nation, rein in consumption or population in a climatologically meaningful way. What's more, though the United States is responsible for a staggering and demographically disproportionate 15 percent of global $CO_2$ emissions as of 2014,[32] there is also the issue of the declining relevance of postindustrial economies from a climatological perspective. China is now the world's largest emitter of $CO_2$ and is rolling out coal-fired power plants in order to add what Ted Fishman describes as "the equivalent of Britain's entire electric output every two years."[33] As sustainability studies scholar Gillen D'Arcy Wood argues, "science has driven this Western and now global project of modernity by a ruthlessly positivist outlook, a 'prideful optimism' that disdains uncertainty,"[34] and it seems increasingly likely that people will soon begin experimenting with strategies designed to recalibrate the Earth's climate systems on a transnational, national, or ad hoc basis. The panoply of strategies currently being floated by scientists and engineers seeking solutions to the problem of climate change include carbon sequestration, cloud farming, fertilizing the oceans with iron powder to promote carbon absorption by plankton, pumping sulfate particles into the stratosphere, and even installing space-based mirrors to deflect solar radiation. Seeking a literary analogue, we might call these ideas "Dorian Gray" solutions for the way they mask and delay a seemingly inevitable day of reckoning.

There are many reasons not to trust in technoscientific "solutions" to anthropogenic warming, including the law of unintended consequences. By looking beyond the dynamo to the coal that powers it, Henry Adams shows us that the problem of the Anthropocene, to riff on W. E. B. Du Bois, is the problem of the externality. According to the *Oxford English Dictionary*, an "externality" is "the condition or fact of being outside another object, or of being an outsider; outward things." Here I want to recruit the term's economic function: it denotes "a side-effect or consequence (of an industrial or commercial activity) which affects other parties without this being reflected in the cost of the goods or services involved; a social cost or benefit. Cf. Spillover." In broad terms, externalities are registered in the differential between the total impact of a behavior or practice and the private transactional costs borne by producers and consumers. Sustainability-minded economists and neoliberals alike have sought market solutions to the problems generated by negative externalities by attempting to monetize their effects either through taxes or through the creation of artificial markets, generating ideas like a carbon tax or the so-called cap-and-trade system, respectively. The goal is to produce, through government action or other strategies (the concept of ecosystems services being the most promising), a market that internalizes what was once external, incorporating the cost of marginal damages into the market price of a good or service and trusting the market to steer itself with an invisible but certain hand back to a sustainable course.

To the extent that the dire predictions of a returning ice age voiced by Adams and his contemporaries remain in circulation at all, they are fodder for climate change deniers eager to cast aspersions on climate science in the present, associating global warming with other debunked Malthusian predictions from ages past. It is not for the accuracy of Henry Adams's climate modeling, however, but for the hermeneutic strategies and aesthetic practices that teach us to read for externalities that theorists of the Anthropocene and those seeking action on climate change would do well to return to his work.

## Notes

1. S. I. Rasool and S. H. Schneider, "Atmospheric Carbon Dioxide and Aerosols: Effects of Large Increases on Global Climate," *Science* 173 (July 1971): 138–41.

2. While the term "nuclear winter" was first used in 1983, the idea that nuclear weapons could bring disastrous climatological consequences in addition to other forms of destruction has a long history in the popular imagination. See especially Samuel Glasstone and Philip Dolan, eds., *The Effects of Nuclear Weapons*, 3rd ed. (Washington, D.C.: U.S. Department of Defense, and Energy Research and Development Administration, 1977); and R. P. Turco et al., "Nuclear Winter: Global Consequences of Multiple Nuclear Explosions," *Science* 222 (1983): 1283–92.

3. Climate change skeptics have devoted significant energy to tracking down evidence of these fears in the 1970s; see especially "1970s Global Cooling Alarmism," *Popular Technology*, February 28, 2013, http://www.populartechnology.net/2013/02/the-1970s-global-cooling-alarmism.html.

4. Ted Cruz, for example, cited the 1975 *Newsweek* article mentioned above as evidence of misguided alarmism and scientific uncertainty in an interview with the *Texas Tribune*. During the campaign Donald Trump invoked a similar logic, arguing that fears of global cooling in the past render current climate science unreliable. For talk radio's Rush Limbaugh and Fox News's Sean Hannity and Lou Dobbs, the myth of a consensus on global cooling in the 1970s supports their denialism and buttresses their claims about the media's supposed liberal bias. For Cruz's interview, see Jay Root and Todd Wiseman, "One-on-One Interview with Ted Cruz," March 24, 2015, https://www.texastribune.org/2015/03/24/livestream-one-on-one-interview-with-ted-cruz. Sources on the cooling myth include John Cook, "What 1970s Science Said About Global Cooling," February 26, 2008, http://www.skepticalscience.com/What-1970s-science-said-about-global-cooling.html; and Doug Struck, "How the 'Global Cooling' Story Came to Be," *Scientific American*, January 10, 2014, http://www.scientificamerican.com/article/how-the-global-cooling-story-came-to-be.

5. See especially Thomas Peterson, William Connolly, and John Fleck, "The Myth of the 1970s Global Cooling Scientific Consensus," *Bulletin of the American Meteorological Society* (September 2008), http://journals.ametsoc.org/doi/pdf/10.1175/2008BAMS2370.1.

6. Henry Adams, *The Education of Henry Adams* (1907; repr., New York: Library of America, 2010), 458 (hereafter cited parenthetically in text).

7. E. M. Forster, *Howards End*, ed. Paul Armstrong (1910; repr., New York: Norton, 1988), 14.

8. Patricia Yaeger, "Editor's Column: Literature in the Ages of Wood, Tallow, Coal, Whale Oil, Gasoline, Atomic Power, and

Other Energy Sources," *PMLA* 126.2 (2011): 305–26, 308.

9. Henry Adams, *A Letter to American Teachers of History* (Washington, D.C.: Furst, 1910), 131–32 (hereafter cited parenthetically in text).

10. Adams seems to have missed the implication of the "closed system" requirement, seeing the dictates of the law of entropy as invalidating the increasing organization of Darwinian evolution and the creative work of the human species.

11. National Oceanic and Atmospheric Administration, "A Paleo Perspective on Global Warming," http://www.ncdc.noaa.gov/paleo/globalwarming/paleobefore.html.

12. Credit for the formulation of the second law is generally attributed jointly to Rudolf Clausius (1850) and William Thomson (1851).

13. John Ruskin, "The Moral of Landscape," in his *Modern Painters* (New York: Wiley, 1889), 2:293–94.

14. Charles F. G. Masterman, *The Condition of England*, 4th ed. (London: Methuen, 1910), 95–96.

15. Frederick Soddy, *Matter and Energy* (Cambridge: Cambridge University Press, 1912), 15 (hereafter cited parenthetically in text).

16. Lewis Mumford, *Technics and Civilization* (New York: Harcourt, Brace, 1934), 157 (hereafter cited parenthetically in text).

17. A. B. Filson Young, *The Complete Motorist: Being an Account of the Evolution and Construction of the Modern Motor-Car, with Notes on the Selection, Use, and Maintenance of the Same, and on the Pleasures of Travel upon the Public Roads*, 5th rev. ed. (New York: McClure, Phillips, 1905), 34.

18. Ralph Waldo Emerson, *Essays: Second Series* (Boston: James Munroe, 1884), 44. The reference to "instant conductors" is from Walt Whitman, "Song of Myself," section 27, in his *The Complete Poems*, ed. Francis Murphy (New York: Penguin Classics, 2005), 91.

19. Charles Baudelaire, *The Painter of Modern Life and Other Essays*, trans. Jonathan Mayne (1863; repr., New York: Phaidon, 1964), 9.

20. Patrick Geddes, *Cities in Evolution: An Introduction to the Town Planning Movement and to the Study of Civics* (London: Williams and Norgate, 1915), 59.

21. Le Corbusier, *Radiant City: Elements of a Doctrine of Urbanism to Be Used as the Basis of Our Machine-Age Civilization* (New York: Orion, 1967).

22. For example, see F. T. Marinetti, "The Founding and Manifesto of Futurism," in *Documents of 20th Century Art: Futurist Manifestos*, ed. Umbro Apollonio, trans. Robert Brain et al. (New York: Viking, 1973), 19–24.

23. E. M. Forster, *The Machine Stops and Other Stories* (New York: Deutsch, 1997), 6.

24. Georges Méliès's film *Le voyage à travers l'impossible* was released by his Star Film production company in the United States as *The Impossible Voyage* and in the United Kingdom as *Whirling the Worlds: The Impossible Voyage*. My quotations are from the English text produced by Star Film's New York office to accompany the film, which can be seen at https://archive.org/details/The_Impossible_Voyage.

25. Jean-Claude Debeir, Jean-Paul Deléage, and Daniel Hémery, *In the Servitude of Power: Energy and Civilisation Through the Ages*, trans. John Barzman (London: Zed, 1990), xii.

26. Leo Marx, *The Machine in the Garden: Technology and the Pastoral Ideal* (Oxford: Oxford University Press, 1964), 191.

27. Will Steffen, Paul J. Crutzen, and John R. McNeill, "The Anthropocene: Are Humans Now Overwhelming the Great Forces of Nature?" *Ambio* 36.8 (2007): 614–21, 616.

28. "IPCC 2013: Summary for Policymakers," in *Climate Change 2013: The Physical Science Basis: Contribution of the Working Group 1 to the Fifth Assessment Report of the Intergovernmental Panel on Climate Change*, ed. T. F. Stocker et al. (New York: Cambridge University Press, 2013), 13.

29. Srinivas Aravamudan, "The Catachronism of Climate Change," *Diacritics* 41.3 (2013): 6–30, 8, 10.

30. Bruno Latour, "Agency at the Time of the Anthropocene," *New Literary History* 45.1

(2014): 1–18, 1. Latour writes of the feeling of temporal belatedness vis-à-vis the concentration of carbon dioxide in the atmosphere: "it seems now very plausible that human actors may arrive *too late* on the stage to have any remedial role.... [T]hrough a complete reversal of Western philosophy's most cherished trope ... nature has unexpectedly taken on that of the active subject.... [T]hrough a surprising inversion of background and foreground, it is *human* history that has become frozen and *natural* history that is taking on a frenetic pace" (11–12).

31. See https://350.org.

32. See https://www.epa.gov/ghgemissions/global-greenhouse-gas-emissions-data#Country.

33. Ted Fishman, *China, Inc.* (New York: Scribner, 2006), 113.

34. Gillen D'Arcy Wood, "What Is Sustainability Studies?," *American Literary History* 24.1 (2011): 1–15, 2.

# 9.
# Stratigraphy and Empire
## *Waiting for the Barbarians*, Reading Under Duress

*Jennifer Wenzel*

*In memory of Anthony Carrigan*

Let us examine what our ultimate legacy is likely to be, the extent to which the human race and its actions are likely to be preserved within geological strata, and thus transported into the far future. This will be an acid test of our ultimate influence, our final footprint on the planet.
—Jan Zalasiewicz, *The Earth After Us*

The mind is prepared by the contemplation of such future revolutions to look for the signs of others, of an analogous nature, in the monuments of the past.
—Charles Lyell, *Principles of Geology*

[T]he question is ultimately not about the laws of history but about who controls the signs of power.
—David Attwell, *J. M. Coetzee: South Africa and the Politics of Writing*

In his first provocation on the Anthropocene, Dipesh Chakrabarty cites Alan Weisman's *The World Without Us* (2007): "Picture a world from which we all suddenly vanished.... Might we have left some faint, enduring mark on the universe?" For Chakrabarty, Weisman's imagined scenario offers a point of entry into problems of time—"a sense of the present that disconnects the future from the past"—that the Anthropocene poses to historiography and epistemology.[1] As a literary critic, I am fascinated by how Weisman's question (and that of geologist Jan Zalasiewicz in the epigraph) resonates with those in J. M. Coetzee's *Waiting for the Barbarians*, a novel about an unnamed magistrate of an unspecified empire. Having spent his career far from the capital, the Magistrate fills his days as an amateur cartographer and archaeologist, excavating the ruins of what he takes to be a fallen empire that preceded the one he serves. In his archaeological pursuits, as in his relationship with the "barbarian girl" tortured by Colonel Joll, a military officer from the capital, the Magistrate is obsessed with tracing the marks of history. He speculates: "Perhaps when I stand on the floor of the [excavated] courthouse, if that is what it is, I stand over the head of a magistrate like myself, another grey-haired servant of Empire who fell in the arena of his authority, face to face at last with the barbarian."[2] Rather than a future world without humans, the Magistrate imagines a series of vanished but identical past worlds—the rise and fall of successive empires—with these strata of history stacked on top of each other like stories in an upside-down skyscraper.

In this chapter, I use *Waiting for the Barbarians* as a matrix for thinking about the Anthropocene, to confront the difficulties of thinking between human and natural history, imperial and geological time. I transpose the historiographical questions Chakrabarty and Zalasiewicz raise about the Anthropocene and those the Magistrate asks about empire into literary (and, more forcefully, political) ones. What would constitute an Anthropocene archive? How and why would we (or those after us) read it? What are the stakes of *desiring* an archive of the human to be left for some deep future after humans? How are meaning and justice connected to the shapes we assume time and history to have? Anthropocene reading, I argue, is reading under duress, with "duress" denoting multiple forms of force at work in human and natural history.

The most common sense of "duress" in contemporary parlance refers to the force of the state: political repression or coercion, (extra)judicial pressure. The iconic scene of reading under duress in *Waiting for the Barbarians* is when the Magistrate-turned-prisoner is interrogated by Joll and made to "read" the inscrutable signs inscribed on poplar slips unearthed in his excavations. The Magistrate offers a false allegory yet speaks truth to power by imputing anti-imperial messages to texts we know he is unable to decipher. An elusive archive, the fragile slips are wrested from their archaeological and hermeneutic context and made to bear meaning in the present. This scene, together with an oblique reference to the 1977 death in detention of Black Consciousness leader Steve Biko, invited readings of the novel as a political allegory for state violence and torture under apartheid. This ripped-from-the-headlines interpretation was at odds with readings that found in the novel's seeming timelessness and geographical unlocatability a universal allegory for empire writ large.[3] Rather than universality, David Attwell identifies in the novel's setting a "strategic refusal of specificity" born of *writing* under duress and "being painfully conscious of one's immediate historical location" (*Face to Face*, 73). The Magistrate's reading of the slips offers a glimpse into the predicament of South African writers under a state of emergency, which Coetzee described in 1986 as "how not to play the game by the rules of the state . . . how to imagine torture and death on one's own terms."[4] Thus the Magistrate's penchant for allegory, where "events are not themselves but stand for other things" (*Waiting*, 40).

I find in *Waiting for the Barbarians* an anticipatory allegory for the Anthropocene that derives partly from how Coetzee confronts reading and writing under duress. Attwell observes that Coetzee engages his historical situation by reframing what is meant by "history." In the Magistrate's failed attempt to write a history of the frontier outpost is an account of how "history emerges as an object in itself . . . objectified as History—thus emerging . . . as the time of crisis, or *kairos*, as against the time of the seasons, or *chronos*."[5] The crisis provoked by Joll's arrival registers as a problem of temporality: "What has made it impossible for us to live in time like fish in water, like birds in air, like children?" the Magistrate asks himself. "It is the fault of Empire! Empire has created the time of history. Empire has located its existence not in the smooth recurrent spinning time of the cycle of the seasons but in the jagged time of rise and fall, of beginning and end, of catastrophe" (*Waiting*, 133). As he contemplates the horrors of imperial time, the Magistrate finds consolation in the passing of the seasons. "How can I accept that disaster has overtaken my life when the world continues to move so tranquilly through its cycles?" (94). Yet he recognizes that nature's temporality is not *only* cyclical; he reads in dried-up lake beds an augury of the outpost's demise, for he knows that the gradual salinization of a nearby lake means that its days as an agricultural settlement are numbered. Over the long term, the increasingly brackish water poses a threat more definite than that imputed to the barbarians. The Magistrate knows better than to assume that human history is distinct from natural history. He confronts problems not unlike those Chakrabarty finds in global warming: how to think between disjunctive temporalities and time scales, how to find meaning in that which resists interpretation. Perhaps the impulse to see the Anthropocene as a "story" or "drama"—a discrete literary artifact, rather than a phenomenon about which many narratives might be fashioned—underwrites the limitations of Chakrabarty's groundbreaking interventions.[6]

The Magistrate's environmental awareness barely registered in the first decades of the novel's reception, precisely because the urgencies of apartheid were understood to belong to the realm of history (and politics) in a way that nature and environmentalism did not.[7] Consider Nadine Gordimer's review of Coetzee's subsequent novel, *Life & Times of Michael K* (1983), where she expresses perplexity about how K's commitment to keeping "the idea of gardening" alive in a time of war could amount to a revolutionary gesture: "Beyond all creeds and moralities, this work of art asserts, there is only one: to keep the earth alive, and only one salvation, the survival that comes from her. . . . Hope is a seed. That's all. That's everything. It's better to live on your knees, planting something?"[8] Even Gordimer's trailing question keeps the idea of gardening alive. Reading *Waiting for the Barbarians* in terms of the Anthropocene, I cannot forget the historical pressures that have shaped the novel's reception and continue to complicate such a reading, including the disconnect between Coetzee's coolly cerebral allegory and the white-hot political turmoil of South Africa, either in the early 1980s or in the present.[9] This reception history is a precedent with its own politics. Hope is a seed; the idea of gardening remains essential. I'm not certain that allegory of any sort is sufficient to the Anthropocene; nonetheless, allegorizing like the Magistrate, I struggle to read and write in a mode adequate to history, answerable to the future.

The Magistrate's slips point to other aspects of reading under duress and other forms of force at work in the Anthropocene. Even before his interrogation, the Magistrate acknowledges that the slips are the product of coercion, excavated by prisoners who would not choose such work under the punishing sun. The tasks of archiving and interpretation depend on the grueling labor of others. In what ways does "duress" describe our own scenes of reading, the material and mineral circumstances that underwrite the labor of interpretation, in situations of relative privilege or precarity? Moreover, "duress" derives from *dūritia*, Latin for "hardness"; it shares this root with "endure." To persist through difficulty is a form of hardening, toughening up. How is reading a form of endurance? As Stephanie LeMenager asks in measuring the ethical and cognitive value (and environmental costs) of slow-reading print against quicksilver digital media, what does the Anthropocene mean for the "traditional arts of duration"?[10]

Duress has everything to do with rocks and fossils: the hardness produced by the Earth's own pressure. If the prisoner offers one model of what it means to read (and write) under duress, what of the geologist or archaeologist, or even the rock, fossil, or artifact? Stratigraphy, the science of layers and layering, is a methodological approach fundamental to the disciplines of geology and archaeology, whose primary objects are produced under conditions of duress: shaped and hardened (or, in the case of archaeology, preserved or disturbed) through geological force. In *The Earth After Us*, Zalasiewicz observes that our planet is distinctive for the formation of its rocky layers from decaying organic matter,[11] but this capacity seems exhausted or gone awry in *Waiting for the Barbarians*. Burdened by a sense of complicity in the present, unable to imagine a future, the Magistrate is a stratigraphic thinker occupied by shards of the past—his historiographical project, his archaeological excavations, images of a tired Earth and improperly buried bodies (*Waiting*, 92, 148). Stratigraphy and empire are joined in his "reading" of the ruins, which reconciles the tension between *kairos* and *chronos*. Replotting imperial time as a cycle is one of the gestures through which the Magistrate disavows complicity with the Empire's violence: there have been magistrates and barbarians before him, and will be afterward. Thinking between empire and the Anthropocene, I want to tease out this politics of the precedent in the shape and scale of time: What are the different implications for justice if one sees history as cyclic or linear, repeated or ruptured, analogous or without precedent?

A turn to precedents also emerges in Chakrabarty's series of interventions. Insisting that capitalism cannot adequately explain the Anthropocene, Chakrabarty's later essays emphasize the cyclic aspect of planetary warming and cooling, so that humans are merely contingent, accidental, and incidental as a cause ("Climate and Capital," 11, 21–22). Situating anthropogenic warming in terms of precedents in the history of Earth and other planets, however, risks forestalling opportunities for political action in this contingency that is our only planetary home. (The U.S. Republicans who acknowledge climate change invoke the glaciation cycle for precisely this reason.) Even if capitalism/colonialism alone cannot account for the Anthropocene, the enduring inequalities they have created necessarily shape how it unfolds and the political dynamics through which humans confront it today. When the Magistrate acknowledges his inability to write a history, he confesses, "There has been something staring me in the face, and still I do not see it" (*Waiting*, 155). What stares the Magistrate and Chakrabarty in the face, arguably, is empire and its inscriptions on bodies and biomes; both would-be historians assume a vision of history that cannot reckon fully with the enduring marks of empire.

Precedent, however, is integral to the very logic of modern geology, as Charles Lyell explains with a startling invocation of empire:

> The annihilation of a multitude of species has already been effected, and will continue ... in a still more rapid ratio, as the colonies of highly-civilized nations spread themselves over unoccupied lands. Yet, if we wield the sword of extermination as we advance, we have no reason to repine at the havoc committed. ... [I]n thus obtaining possession of the earth by conquest, and defending our acquisitions by force, we exercise no exclusive prerogative. ... The most insignificant and diminutive species, whether in the animal or vegetable kingdom, have each slaughtered their thousands, as they disseminated themselves over the globe.[12]

Lyell finds that extinctions have happened before, as part of the conflict inherent to species encounters. He borrows the language of colonial conquest to describe the effects of humans on plant and animal species. This passage cannot be read as justifying colonial genocide against humans, but only because the phrase "unoccupied lands" renders the Earth *terra nullius*. Later in *Principles of Geology*, Lyell acknowledges that "few future events are more certain than the speedy extermination of the Indians of North America and the savages of New Holland in the course of a few centuries, when these tribes will be remembered only in poetry and tradition."[13] In a troubling double naturalization, Lyell reduces colonial expansion to a simple story of humans exterminating plant and animal species; moreover, this process of destruction is what every species seeks to do to every other. The human war against nature is itself part of nature, and the prospect of colonial genocide offers but a "faint image" of the precedent of universal interspecies competition (3:162). (The European conquest of Native Americans was indeed embedded in nature, given the role of Old World microbes as conquistadores.) The heading running across these pages reads: "POWER OF EXTERMINATING SPECIES / NO PREROGATIVE OF MAN." Empire and extinction are brought into the realm of the natural and the cyclic: colonial conquest is both a metaphor for extinction and a cause and accelerant of it. (The hinge is the "as" in "as the colonies of highly-civilized nations spread": its initial temporal sense of "while" is supplanted by the analogical sense of "just as" in the sentences that follow.) This rhetorical move is implicated in the politics of the precedent—by which I mean both Lyell's use of other species extinctions as a precedent to naturalize the effects of humans, and Lyell himself as a precedent for discourse on the Anthropocene. Lyell was among the nineteenth-century scientists who asserted the need to demarcate the geological epoch that became known as the Holocene, which began nearly twelve thousand years ago and is distinguished in part by the rise of *Homo sapiens,* whose spread across the planet Zalasiewicz dubs "human empire" (*Earth After Us*, 1).

The obvious question to ask of Lyell's remarks on species extermination is the one that climate justice activists ask about Anthropocene talk of humans as a species: Who is the "we"? Who wields the "sword of extermination": "we" humans or "we" Europeans? This slippage is crucial to Lyell's naturalizations. We can borrow Lyell's logic and read against the grain to reconsider universalist approaches to the Anthropocene, which argue that extinctions, global warming, and other changes are "NO PREROGATIVE OF" the West, capitalism, or colonialism, but part of a longer human (and planetary) story. From this perspective, the citation of precedents for our current predicament depoliticizes the present.

There is also a resonance between Lyell's naturalization of human violence and the curious verbs that Chakrabarty uses in his first climate essay: humans have "fallen into," "stumbled," "slid into" the Anthropocene ("Climate of History," 216, 218, 219, 217). This repeated locution connotes a hapless accident, a banana-peel mishap at a planetary scale—which brings to mind John Robert Seeley's notorious quip that the British "seem, as it were, to have conquered and peopled half the world in a fit of absence of mind."[14] This absent-minded Anthropocene makes plain the need for a more supple account of the relationship between intention in human history and causation in geophysical history, a relationship that itself has a history. At some point—say, around the time that Exxon decided to bury what its scientists had discovered about the climate consequences of fossil fuels—the Anthropocene shifted to something less like an accident and more like an intentional act or conscious choice, at least among those with the knowledge and power to act otherwise.[15] Exxon's shift has not been proposed as a date for the onset of the Anthropocene; imagine for a moment, however, that the designation of a new geological epoch depended not (only) on a localized stratigraphic signature reflecting changes at a planetary scale, but on a single human signature with implications for all life on Earth: a signing off on an understanding of what humans have wrought with fossil fuels, which consigned humans and nonhumans alike to a different world.

What made the past legible as precedent in stratigraphy was the insight that depth under the Earth's surface correlated with historical time. This is the law of superposition: the layers of rock or archaeological ruins closest to the Earth's surface are the most recent; deepest is oldest. Because the logic of superposition has acquired the status of common sense (for reasons I discuss below), we tend to forget how it transformed the way history was understood to be inscribed in layers of sediment. (In other words, a new understanding of the way rocks and ruins were produced or "written" under duress shaped the way they were to be interpreted or "read": this insight has interesting implications for the relation between literary production and reception in the Anthropocene, a question to which I return below.) For scholars in postcolonial or African studies, moreover, the law of superposition has a troubling air of mastery and hierarchy about it. The problem is not that "superposition" sounds like "superiority," but that geological distance beneath the Earth's surface actually works in the way that European imperialists falsely posited about geographical distance across the Earth's surface.

Superposition assumes the identity of space and time: the farther down you go, the further back in time you are. Imperialism's civilizational impulses also assumed the identity of space and time by conflating geographic distance or cultural difference with temporality: movement away from Europe was imagined as travel back in time. This "denial of coevalness," in Johannes Fabian's classic formulation, was the refusal of Europeans to accept that the peoples they encountered occupied the same moment in time.[16] Modern geology and archaeology emerged at a moment of European exploration and imperial expansion. From this perspective, it becomes possible to imagine a late eighteenth- and nineteenth-century cartographical project in three dimensions, with an implicit fourth dimension and temporal axis: on the one hand, the stratigraphic logic of geological superposition (vertical distance as epochal time); on the other, the imperialist logic of geographical expansion (horizontal distance as developmental time). In the case of movement below the Earth's surface, superposition was a conceptual breakthrough that facilitated modern stratigraphic knowledge. In the case of movement across the Earth's surface, the denial of coevalness provided a justification for colonizing supposedly primitive peoples. Yet it would be inadequate to see one as legitimate science and the other as unfortunate ethnocentrism. In each case, nature and politics intertwine; even if the ethnocentrism of imperialist space-time is now readily apparent, the chronological insight of stratigraphy was once scandalous for its displacement of anthropocentric biblical history. This is the danger of metaphorizing nature in terms of intrahuman politics, and vice versa: the pitfalls of geological consciousness (troping on Frantz Fanon's anatomy of the "pitfalls of national consciousness") or what Louise Green calls "the alibi of geology."[17]

Stratigraphy can be understood as a form of reading under duress, with superposition as its underlying logic—a logic more fundamental to literary interpretation than we might expect. An Anthropocene perspective can recast recent debates about symptomatic versus surface reading. In their framing of this debate, Stephen Best and Sharon Marcus assert that symptomatic reading, whether modeled on psychoanalysis or Marxism, posits literary meaning as "hidden, repressed, deep, and in need of detection and disclosure by an interpreter." Symptomatic reading attends to absences, latencies, and depths, driven by a concern with "what those absences mean, *what forces create them.*"[18] One question that goes all but unasked in this debate is how the "forces" at work in psychic repression compare to those at work in geological sedimentation and fossilization.[19] After all, psychoanalysis and the stratigraphic sciences of geology and archaeology are fundamentally concerned with depths, whether subconscious or subterranean. This isomorphism is neither accident nor coincidence. Sigmund Freud was fascinated by archaeology during a "golden age of archaeological discovery" that saw high-profile excavations in Greece, Egypt, and Sumeria.[20] An avid collector of antiquities, Freud drew on the stratigraphic aspects of archaeology (and, to a lesser extent, geology) as an analogy or metaphor for the unconscious.

For Freud, the psychoanalyst is an archaeologist, digging down through sedimented layers of the mind and memory to seek the origins of psychopathology, guided by stratigraphy's law of superposition: deepest is oldest. The burial of Pompeii emblematizes the forces at work in psychic repression: "There is, in fact, no better analogy for repression, by which something in the mind is at once made inaccessible and preserved, than burial of the sort to which Pompeii fell a victim and from which it could emerge once more through the work of spades" (Freud qtd. in Bowie, *Freud*, 19). Archaeology, according to Malcolm Bowie, offers to psychoanalysis a "picture of time made perfectly legible in layered deposits of matter," "a dream of unitary and unidirectional knowledge" (23, 25). Like the archaeologist with his "picks, shovels, and spades," Freud understood the psychoanalyst as having the power to bring the buried past to light, to make the stones speak (qtd. in Bowie, *Freud*, 19).

Thus, symptomatic reading *is* stratigraphic reading. Reading under duress, as I envision it, recognizes the relationships among these forces: how psychic repression (imagined as a form of burial) is modeled after geological sedimentation and differs from the forces at work in political repression. Current attempts to make sense of what geologic time scales mean for literature will have to grapple with the largely unrecognized geological assumptions *already* at work in reading practices that have long dominated our discipline. Like the forces at work in psychic repression, geological forces create layers of rock that overdetermine our scenes of reading: in them, we read what life on Earth was like in epochs past; through them, we inhabit modernity (libraries, databases, electric grids, delivery trucks, asphalt, iPads) and its uncertain futures. What does it mean to turn away from what lies below the surface at the moment we are becoming aware that the near future will be shaped by a past that persists in atmospheric and oceanic $CO_2$ as well as in the rock record?

Giving up the hermeneutics of suspicion and "stay[ing] close to our objects of study," according to Best and Marcus, can yield more "*accurate* accounts of surfaces" ("Surface Reading," 15, 18, emphasis added). This assertion of accuracy as a commonsensical interpretive value is challenged by Carolyn Lesjak's distinction between surface reading's matter-of-fact "knowing what we see" and a dialectical concern for "seeing what we know" (a phrase she borrows from Eve Sedgwick) in which "the status of the real is an interpretive problem rather than a given."[21] Assessing the import of surface reading for South African literary studies, Sarah Nuttall construes surface as skin and depth as mine, both of which are inevitably historically freighted, so that the surface is a "fundamentally *generative force* capable of producing effects of its own"; it is thus (along with depth and symptom) "a place to think from" ("Rise," 410, 417). Pushing Lesjak's "Reading Dialectically" in a geological direction, we might read modern geology as having emerged from a crisis of "seeing what we know" and "knowing what we see" with regard to geological phenomena upon and beneath the Earth's surface. What does it mean to be "accurate" or faithful to the "real" if the temporal scale of observation shifts from the human lifespan to the geological epoch? Such a shift is under way in Anthropocene discourse, which revises our understanding of basic facts regarding the causes and consequences of modernity.

Freud's *Civilization and Its Discontents* offers a thought experiment that complicates the notion of surfaces as being simply there to "just read" by playing tricks with time at an archaeological scale (Best and Marcus, "Surface Reading," 12). In another convergence of stratigraphy and empire, Freud imagines a view of Rome with all its successive instantiations and historical layers visible at one time:

> Now let us make the fantastic supposition that Rome were not a human dwelling-place, but a mental entity with just as long and varied a past history: that is, in which nothing once constructed had perished, and all the earlier stages of development had survived alongside the latest. . . .
>
> And the observer would need merely to shift the focus of his eyes, perhaps, or change his position, in order to call up a view of either the one or the other.[22]

In Freud's archaeological fantasy, Italian, Roman, and Etruscan edifices that have successively occupied the same place are seen to coexist in the same place—at the same time. This fantasy transforms stratigraphy's layered past to simultaneity; it rearranges precedence as coexistence, historical succession as timelessness. We could think of it as a fantasy of a particular *kind* of surface reading, or as a horizontal counterpart to the vertical image of magistrate standing atop magistrate, or as a vivid and unwitting instantiation of the effect of using vast spans of time as a framework for events separated by decades, centuries, or millennia. The rise and fall of these edifices (and their empires) might as well be simultaneous when considered at the scale of the 4.5 billion years of Earth's history: "Even spatially heterogeneous and diachronous . . . changes appear instantaneous when viewed millions of years after the event, especially as time-lags often fall within the error of the dating techniques" (Lewis and Maslin, "Defining," 173). This statement echoes (in technical language) a passage in Lyell's *Principles of Geology* that construes such time-scalar differences in terms of literary mode and genre: events in the history of a nation that would read as realistic and credible in a narrative spanning two thousand years would "immediately assume the air of a romance" full of wild improbabilities and supernatural incidents were they plotted over a single century.[23]

Freud sketches his Roman fantasy to argue for the *permanence* of psychic traces in the unconscious: nothing perishes in this impossible vision of rise without fall. Consider the resonances between Freud's Rome and Achille Mbembe's Johannesburg:

> With the end of legally sanctioned segregation, Johannesburg is nowadays [2008] a metropolis increasingly forced to construct itself out of heterogeneous fragments and fortuitous juxtapositions of images, memories, citations, and allusions drawn from its splintered histories. . . . While bearing witness to a demand that the past be forgotten, this architecture asks the spectator to forget that it is itself a sign of forgetting. . . . The rupture between the racist past and the metropolitan present, between here and there and between memories of things and events, renders possible the production of new figural forms and calls into play a chain of substitutions.[24]

Mbembe finds a temporal folding at work in Johannesburg's multiracial collage of heterotemporal fragments, which he reads in terms of hysteria and conjuration, constraint and possibility. White nostalgia for the racially ordered city generates what Mbembe calls "fantasies" and "hallucinations" that implicitly link the topography of Johannesburg back to Freud's Rome and its "chain of substitutions": "The unconscious can easily accommodate the survival of archaic formations beside others that have supplanted them, even on the same site" (Hubert Damisch qtd. in Mbembe, "Aesthetics," 63). "The unconscious of a city," Mbembe concludes, "is made up of different layers of historical time superimposed on one another, different architectural strata or residues from earlier times" (64).

I turn to contemporary Johannesburg not merely as an instantiation of Freud's fantasy, but also for Mbembe's attention to the historical present as an uneven site of improvisation and imagination. Johannesburg elicits not only racist hallucinations of an undisturbed past, but also consumerist energies so pervasive that "desire is inculcated even in those who have nothing to buy" (Mbembe, "Aesthetics," 64). It is in places like this that many present and near-future humans will ride out the Anthropocene, with little in the way of material resources or infrastructure to cushion the blows of storms and droughts to come. The point is not to cast the shadow of sober reality over allegorical flights of fancy; for Mbembe, play, figuration, and chains of substitution become possible (and necessary) in the palimpsest cityscape that is Jozi. Rather, the investment in the present and in the present tense interests me as an ethical and political counterweight to the future projections, counterfactual speculations, and analogies across vast spans of time assembled in this chapter—Weisman's world without us, Zalasiewicz's archive in the rocks, the Magistrate's excavations, Lyell's clash of species empires and romance versus realism thought experiment, Freud's Roman fantasy.

Such temporally promiscuous imagining might be an emergent genre convention of writing the Anthropocene; perhaps "Anthropocene" names a frantic search for precedent and precipitate, in the distant past or in the imagined future, anywhere but here and now. At the end of Lyell's chapter on species extinction is a glimmer of Anthropocene discourse in this mode: "[T]hroughout myriads of future ages, [these causes of extinction] must work an entire change in the state of the organic creation, not merely on the continents and islands, where the power of man is chiefly exerted, but in the great ocean, where his control is almost unknown. The mind is prepared by the contemplation of such future revolutions to look for the signs of others, of an analogous nature, in the monuments of the past" (*Principles of Geology*, 2:156–57). Lyell's vision of the planet in the wake of this "entire change" is paradigmatic not only as an anticipatory account of the Anthropocene, but also as a vision straddling the far future and remote past. He forges precedents in a temporally multivalent sense: the act of imagining a future that is difficult to imagine serves as a cognitive precedent, preparing the mind to search for historical precedents in geological strata. The narrative premise of Zalasiewicz's *The Earth After Us* works similarly: he imagines the arrival of inquisitive aliens one hundred million years in the future in order to lend the appeal of speculative fiction to his account of how geologists today know what they know about the distant past. "If *you* desire immortality for some aspect of your own personal sojourn on Earth ... to adorn some museum of the far future, *read on*," he exhorts in the book's introduction (1, second emphasis added).

Anthropocene reading, like all reading, has many forms and motive forces; one can hardly begrudge Zalasiewicz his gambit of borrowing sci-fi tropes to bring geological understanding to a general audience. But the desire he names is not my desire; I do not read so that some other species might read about me in the deep future. As Mark Twain might say, the reports of our extinction have been greatly exaggerated. As Ian Baucom writes, "extinction is not the sole copula between the subject and this situation"; within the present moment are "multiple scales, orders, and classes of time ... and multiple corresponding orientations to the possibility of the (just) future fashioning of those times."[25] Coetzee's Magistrate, too, is cognizant of being beset by "dreams of ends: dreams not of how to live but of how to die" (*Waiting*, 133). Nonetheless, two scenes of imagined burial in *Waiting for the Barbarians* register an orientation toward the future as an orientation toward human survival and justice. In a moment of disgust with the barbarians, the Magistrate resolves:

> It would be best if this obscure chapter in the history of the world were terminated at once, if these ugly people were obliterated from the face of the earth and we swore to make a new start.... It would cost little to march them out into the desert ... to have them dig ... a pit large enough for all of them to lie in ... and, leaving them buried there forever and forever, to come back to the walled town full of new intentions. But that will not be my way. The new men of Empire are the ones who believe in fresh starts, new chapters, clean pages. (24)

This image of forcing the barbarians to bury themselves takes to its logical conclusion Lyell's discussion of colonial genocide as an inevitable event within a longer history of species extinction. Leaving aside the question of genocide, where the Magistrate and Lyell differ most crucially from the "new men of

Empire" is in rejecting the catastrophist idea that anything can remain buried "forever and forever" and that new futures can be severed from the past. Even after colonialism would have wiped out Indigenous peoples, Lyell expected vanished human tribes to live on "in poetry and tradition." History will out; the deep past remains legible to those who can read it. The Magistrate waits, not so much for the barbarians, but for the historians. Recognizing his inability to write "a record of settlement to be left for posterity buried under the walls of our town . . . as a gesture to the people who inhabited the ruins in the desert," he coats the poplar slips in linseed oil and resolves to rebury them when the weather improves, as fragments of more interest to the future than anything he might write (*Waiting*, 155).

Burying the slips rather than burying the barbarians would be a commitment to the future that is also a commitment to history. Although it might be understood to leapfrog the present, so that the people the Magistrate imagines will "one day . . . come scratching around in the ruins" might as well be Zalasiewicz's alien paleoecologists, the Magistrate's imagined gesture differs crucially from Zalasiewicz's promise of immortality in that he is content not to be inscribed in (or to inscribe) the archive they will unearth. *Waiting for the Barbarians* stands in for the Magistrate's unwritten history; I have come scratching around in it to see what it might portend about the way we read *now*. Reading under duress also names a (fortuitous) form of constraint that prevents allegory from being friction- or gravity-free, able to loose the ties that bind a text to its anguished scene of writing. If there are many Anthropocenes, some of them are African. In *Waiting for the Barbarians*, to "leave [one's] mark" is equally to wound as to write: the desire to write oneself into an archive can be its own kind of violence (112).

Perhaps I do violence to Coetzee's novel by reading it as an anticipatory allegory for the Anthropocene. If so, that violence is of a particular kind: the Benjaminian *Jetztzeit* or "now of a particular recognizability" that wrenches historical moments out of the smooth, undeformed logic of succession, historicism's one damned thing after another.[26] The Earth's atmosphere, hydrosphere, and lithosphere have been inscribed as an archive of greenhouse gases that we are only beginning to learn to read. A literary history for the Anthropocene that functions as a "force of nature," as Baucom hopes,[27] will not abide a neat stratigraphic logic, which aligns with the kind of thinking Benjamin derided as historicism, where the past is safely past, neatly buried under the present in smooth and legible layers. Rather, it will acknowledge unconformities, to borrow Eric Gidal's argument that "periods of dislocation and erosion in the discontinuous strata of sedimentation" in the realm of geology have their literary counterparts in "periods of economic displacement and environmental transformation in the irregular divisions assembled on the surface of the page."[28]

I prefer Benjamin's weak messianic hope for the possibility latent in every present moment over a geological expectation of judgment deferred to some far future. Perhaps the only archive susceptible to politics is the one in the atmosphere and the oceans, rather than the one written in stone. There is meaning to be found in geology but little justice in the rocks. That's the rupture of modern geology, wresting the strata and the forces that shaped them out of the realm of divine will or transcendence. The way we humans read Anthropocene archives now—whether those in water, air, and rock or those in the slips we call literature—matters in these times of duress.

## Notes

1. Dipesh Chakrabarty, "The Climate of History: Four Theses," *Critical Inquiry* 35.2 (2009): 197–222, 197 (hereafter cited parenthetically in text).

2. J. M. Coetzee, *Waiting for the Barbarians* (New York: Penguin, 1980), 15–16 (hereafter cited parenthetically in text).

3. The novel's setting includes flora and fauna of the Southern Hemisphere and seasons of the Northern Hemisphere. Coetzee's notebooks indicate that he imagined the landscape as an amalgam of the Karoo desert and the Mongolian steppe. See David Attwell, *Face to Face with Time: J. M. Coetzee and the Life of Writing* (New York: Viking, 2015), 114, 116 (hereafter cited parenthetically in text).

4. J. M. Coetzee, "Into the Dark Chamber: The Writer and the South African State," in his *Doubling the Point: Essays and Interviews*, ed. David Attwell (Cambridge: Harvard University Press, 1992), 366.

5. David Attwell, *J. M. Coetzee: South Africa and the Politics of Writing* (Berkeley: University of California Press, 1993), 72, 75.

6. Dipesh Chakrabarty, "Climate and Capital: On Conjoined Histories," *Critical Inquiry* 41 (2014): 1–23, 15, 20, 21 (hereafter cited parenthetically in text).

7. The Magistrate also observes the ways that nature is weaponized in the conflict between the Empire and the barbarians.

8. Nadine Gordimer, "The Idea of Gardening," *New York Review of Books*, February 2, 1984.

9. Sarah Nuttall frames this structural and generational problem in terms relevant to my argument: "[L]arge bodies of literature by both young black and white South African writers pass by almost unremarked and unnoticed as critics focus compulsively on the fiction of Coetzee, Gordimer, [Damon] Galgut, and sometimes Zakes Mda. An emerging and implicit critique is slowly gathering steam in literary studies in South Africa of analyses that exclusively take a certain kind of self-conscious, psychoanalytically driven literary genre for granted as the key object of study.... Rather, it might be useful for literature departments to embrace those kinds of literature that do not always need to be decoded in order to be understood in the way that 'the political unconscious' requires." Nuttall, "The Rise of the Surface: Theorising New Directions for Reading and Criticism in South Africa," in *Print, Text, and Book Cultures in South Africa*, ed. Andrew van der Vlies (Johannesburg: University of

the Witwatersrand Press, 2012), 408–21, 417 (hereafter cited parenthetically in text).

10. Stephanie LeMenager, *Living Oil: Petroleum Culture in the American Century* (Oxford: Oxford University Press, 2014), 122–23.

11. Jan Zalasiewicz, *The Earth After Us: What Legacy Will Humans Leave in the Rocks?* (Oxford: Oxford University Press, 2008), 22–25 (hereafter cited parenthetically in text).

12. Charles Lyell, *Principles of Geology: Being an Attempt to Explain the Former Changes of the Earth's Surface, by Reference to Causes Now in Operation* (London: John Murray, 1832), 2:156.

13. Lyell, *Principles of Geology: Being an Inquiry How Far the Former Changes of the Earth's Surface Are Referable to Causes Now in Operation*, 4th ed. (London: John Murray, 1835), 3:163 (hereafter cited parenthetically in text).

14. J. R. Seeley, *The Expansion of England* (London: Macmillan, 1883), 8. Seeley tends to be quoted without including the phrase "and peopled"; read with Lyell and with Lewis and Maslin's Orbis hypothesis, this peopling was also an epochal unpeopling and eradication of other species. See Simon L. Lewis and Mark A. Maslin, "Defining the Anthropocene," *Nature* 519 (March 2015): 171–80 (hereafter cited parenthetically in text).

15. See Neela Banerjee, Lisa Song, and David Hasemyer, "Exxon: The Road Not Taken," *Inside Climate News*, September 16, 2015, https://insideclimatenews.org/content/Exxon-The-Road-Not-Taken.

16. See Johannes Fabian, *Time and the Other: How Anthropology Makes Its Object* (New York: Columbia University Press, 1983).

17. Frantz Fanon, "The Pitfalls of National Consciousness," in his *The Wretched of the Earth* (New York: Grove, 1963), 148–205; Louise Green, "Living in the Subjunctive: Africa in the Anthropocene," paper presented at "Literature in the Age of the Anthropocene: WiSER," University of the Witwatersrand, Johannesburg, November 10–11, 2015.

18. Stephen Best and Sharon Marcus, "Surface Reading: An Introduction," *Representations* 108.1 (2009): 1–21, 1, 3, emphasis added (hereafter cited parenthetically in text).

19. Mary Thomas Crane, "Surface, Depth, and the Spatial Imaginary: A Cognitive Reading of the Political Unconscious," *Representations* 108.1 (2009): 76–97, esp. 90.

20. Malcolm Bowie, *Freud, Proust, and Lacan* (Cambridge: Cambridge University Press, 1987), 18 (hereafter cited parenthetically in text).

21. Carolyn Lesjak, "Reading Dialectically," *Criticism* 55.2 (2013): 233–77, 253–54.

22. Sigmund Freud, *Civilization and Its Discontents*, trans. Joan Riviere (London: Hogarth, 1930), 6.

23. Charles Lyell, *Principles of Geology; or, The Modern Changes of the Earth and Its Inhabitants*, 6th ed. (London: John Murray, 1840), 1:116; discussed in Morgan Vanek, "Coincidence, the Anthropocene, and the Eighteenth Century," paper presented at the annual meeting of the American Comparative Literature Association, Seattle, March 27, 2015.

24. Achille Mbembe, "The Aesthetics of Superfluity," in *Johannesburg: The Elusive Metropolis*, ed. Sarah Nuttall and Achille Mbembe (Durham: Duke University Press, 2008), 59, 62, 63 (hereafter cited parenthetically in text).

25. Ian Baucom, "History 4°: Postcolonial Method and Anthropocene Time," *Cambridge Journal of Postcolonial Literary Inquiry* 1.1 (2014): 123–42, 141–42.

26. Walter Benjamin, *The Arcades Project*, trans. Howard Eiland and Kevin McLaughlin (Cambridge: Harvard University Press, 1999), 463. Elsewhere, Benjamin echoes Freud's archaeological account of the unconscious: "He who seeks to approach his own buried past must conduct himself like a man digging. . . . For the 'matter itself' is no more than the strata which yield their long-sought secrets only to the most meticulous investigation. . . . [G]enuine memory must therefore yield an image of the person who remembers, in the same way a good archaeological report not only informs us about the strata from which its findings originate, but also gives an account of the strata which first had to be broken through." Benjamin,

"Excavation and Memory," in his *Selected Writings*, vol. 2: 1927–34, ed. Michael W. Jennings, Howard Eiland, and Gary Smith (Cambridge: Belknap, 1999), 576.

27. Ian Baucom, "The Human Shore: Postcolonial Studies in an Age of Natural Science," *History in the Present* 2.1 (2012): 1–23, 12.

28. Eric Gidal, *Ossianic Unconformities: Bardic Poetry in the Industrial Age* (Charlottesville: University of Virginia Press, 2015), 15.

# 10.
# Reading Vulnerably
## Indigeneity and the Scale of Harm

*Matt Hooley*

**Corpse Whale**

The title of d. g. nanouk okpik's *Corpse Whale* (2012) names a colonial necropolitical fantasy of vulnerability and a problem of reading. "Corpse whale" is an etymological transposition of the English name for a species of whale, the narwhal (from the Old Norse *na/r* "corpse" + *hvalr* "whale"). Narwhals are not terribly large, and they have pale gray, mottled skin. In the summer they float ("log") just below the surface of Arctic waters, a posture non-Indigenous observers have "likened to that of a drowned human corpse."[1] Narwhals are also called "unicorn whales," because their tusks were sold for centuries by Europeans as unicorn horns. Narwhal tusks were thought to guard against poisoning, though narwhal meat was believed to have the opposite effect, as the sixteenth-century geographer Gerardus Mercator postulated: "Anyone who eats its flesh dies immediately."[2] Narwhals were regarded as savage: in 1555 Olaus Magnus described a "sea-monster that has in its brow a very large horn wherewith it can pierce and wreck vessels and destroy many men."[3] They were also considered regal: narwhal tusks were "incorporated into the British coat of arms by James I in 1603, and in 1671 Christian V became the first Danish king to be crowned in a coronation chair made entirely of narwhal tusks" (Lopez, *Arctic Dreams*, 143).

Despite being overburdened with European mythic meanings, in 1986 Barry Lopez admitted, "we know more about the rings of Saturn than we know about the narwhal" (*Arctic Dreams*, 128). Narwhals are a problem of meaning mapped as both obscurity ("no large mammal in the Northern Hemisphere comes as close as the narwhal to having its very existence doubted"; 128) and excess. Lopez noted that narwhals are immersed in a "different hierarchy of senses than we are accustomed to," a "three dimensional acoustic space" so rich that it frustrates human apprehension: "A later study . . . found narwhals 'extremely loquacious underwater,' and noted that tape recordings were '[so] saturated with acoustic signals of highly variable duration and frequency composition'" that "much of the narwhal's acoustically related behavior 'remains a matter of conjecture'" (139–40). The narwhal's world is one whose inaccessibility does not diminish our desire to attach to it, and this is no less so in an era of global climate change, when Arctic ecosystems are characterized as too vulnerable to endure much longer: rarefied spaces of gorgeous and terrible disappearance. Today, warming Arctic waters disrupt narwhals' migration and feeding, and their exquisite soundscapes are shattered by seismic air-gun oil deposit surveys—explosions that echo for thousands of miles.[4]

The narwhal's name and cultural history suggest that U.S. and European attachments to it have always been deathly, or necropolitical. That is, death has been read onto and out from the narwhal. In this chapter, I call that action of ascribing and scaling out necropolitical meaning "reading vulnerability." Because that practice structures U.S. and European environmental politics, I argue that settler environmentalism is not (as it portrays itself) a recuperative system of political sensibility, but a way of consolidating power and erasing difference. As an alternative, I propose a way of reading vulnerably that considers experiences and epistemes that throw into relief the limits of political and cultural logics to address in their own terms the ecological crises that they themselves precipitated. To do this, I center okpik's *Corpse Whale*—a volume rooted in the Inupiaq language, environments, and sociality, which thwarts readers' attempts to ascribe or abstract global environmental meaning from it. Reading *Corpse Whale* provincializes mainstream ecopolitics and retheorizes disciplinary vulnerability as a generative practice of epistemic encounter and exchange.

In the context of the Anthropocene, okpik's volume enables me to challenge the specific tendency of theories of global environmental crisis to constitute themselves through the erasure of Indigenous people and knowledge. For example, Stephen Pax Leonard demonstrates that nostalgia for the narwhal's and other species' vulnerability easily extends to the Indigenous people who have coexisted with them for millennia. He observes that receding sea ice and decimated game populations have "left the Inughuit believing that their current settlements will not be here in 15 years' time. [And] if the Inughuit are forced to leave their ancient homeland, it is likely that the language of these Arctic hunters will disappear. With it, their already endangered ancient spoken traditions—a rich depository of indigenous cultural knowledge about how they relate to the land, sea and ice, bound up in stories, myths and folklore—will also be lost."[5] Lumping Indigenous people and ecosystems into narratives of inevitable demise is a familiar colonial habit that okpik challenges in her poem "Her/My Arctic: Corpse Whale":

> It comes back to the Inuit me:
> images     in the mirror are closer     than they appear
> . . . . . . . . . . . . . . . . . . . . . . . . .
> While she's/I'm paddling              another floating corpse,
> a spotted human pelt                          a narwhal is passing
>                     a turquoise iceberg.
> Of plucked         bones of ivory   with spiral blood    stained ribbons
>
> reduced to a single        tusk. She/I pass/es     and keep/s paddling.[6]

Throughout *Corpse Whale*, okpik's subject is dispersed across a forward slash. This is not a writerly ambivalence, but a challenge to reading.[7] It inaugurates a poetics of suspension that the volume refuses to resolve: the moment we pause to figure out how to read "she's/I'm paddling," it opens, distending the scene of action and the coordinates of space, time, and meaning we take that moment to issue. The instant of recognition ("another floating corpse, / a spotted human pelt") does not gather the subject and the whale together in tragic vulnerability (a "passing," a becoming past) but warps into sonic exchange, a passing between: "She/I pass/es and keep/s paddling." Readers are left out of that exchange (it transpires in a blank space) so we cannot say it is redemptive. But this is the point: that a moment activated in transformation might defy readerly attempts to distribute ideological meaning onto it or to abstract political meaning from it. This intimate transformation is erased when such a richly inscrutable scene is remapped as a disappearing world.

The Anthropocene calls for new accounts of power and history given that "mankind [is] . . . an environmental force" capable of altering Earth systems.[8]

What I call "reading vulnerability" points out that these accounts remap Indigenous communities and ecosystems as fragile or disappearing in order to conceive and express their political stakes. Vulnerability is an ascription—a rewriting of complex, even inscrutable, experiences of environmental harm as readable. And it is the ascription of disappearance specifically that yokes these new accounts of human power to Indigenous death—something I use Achille Mbembe's necropolitics to examine.

To understand what kinds of social and intellectual life a construct like "disappearing world" disappears and what it makes legible only as disappearance, I consider the scalar function of readable vulnerability in the Anthropocene. In "The Climate of History" Dipesh Chakrabarty suggests we "scale up our imagination of the human" to address "a shared catastrophe that we have all fallen into."[9] It has been pointed out that scaling up the human makes it easy to erase differences of responsibility and impact. I am also concerned that it is via a stabilized sense of what vulnerability is and what it means that a planetary concept of environmental history becomes thinkable. Suggesting the stakes of his argument, Chakrabarty warns, "there are no lifeboats for the rich" ("Climate," 221). For Chakrabarty, vulnerability is politically legible *and* scalable: it is what connects the metaphoric lifeboat and the materially precarious planet.

A scalable vulnerability disappears different Indigenous experiences of environmental harm and uses Indigenous disappearance as a way of making global environmental harm legible. Simon Lewis and Mark Maslin, for instance, date the Anthropocene to the death of fifty million Indigenous people in the Americas and a corresponding dip in atmospheric carbon dioxide—an event, they argue, that is stratigraphically legible and that links climate change to global empire.[10] While identifying this role of empire is essential, Lewis and Maslin do not escape a necropolitical construction of vulnerability. Their global environmental history not only assumes the readability of Indigenous disappearance, but it makes Indigenous people and knowledge scientifically legible only in or as disappearance.

In the final section of this chapter, I argue that reading global climate change does not require scaling up to the geologic or planetary but rather revising the relationship among vulnerability, agency, and environment. To read vulnerability differently, "to think vulnerability and agency together," as Judith Butler puts it,[11] we should not accept that we know how to read vulnerability or what reading it means for "us." Shifting toward reading vulnerably begins by taking seriously the epistemic value of experiences of vulnerability that Anna Tsing terms "nonscalable"—experiences of vulnerability that operate both as "impediments" to existing and unsatisfactory disciplinary reading practices and as "incitements" to make those reading practices themselves vulnerable in new ways.[12]

In "Her/My Arctic: Corpse Whale" okpik blurs the signs and gestures that could be scalable or erasable. The poem's subject unravels across multispecies affinities, from the narwhal, "our carnage fuel oil wicks in lighted igloos," to

>   Her/my flouncing caribou      in dark moonlight are dodging      Bush laws.
>
>   Her/my Malamute     trots in Arctic circles
>   before the midnight                                           storm.
>   Her/my ringed seal barks     couplets of foreshadows      in an oval tasting
>
>   room
>   with white columns and musty     yellowed law books. (66)

The subject does not gather against species vulnerability but disperses itself across a landscape of shared interest and agency, which theories of species vulnerability in the Anthropocene too often map over. Thus, this poem antagonizes existing protocols of Anthropocene reading. It invites a mode of reading vulnerably, a nonscalable relationship between agency and vulnerability that generatively exceeds disciplinary apprehension:

>                    ... she/I try/ies pushing,
>
>   pushing, and shoving                    the sinew back    into the threaded
>                      bones of the land. (66)

**Reading Vulnerability**

"Reading vulnerability" refers to the way vulnerability to environmental harm is understood to be legible and scalable. Like others, I am interested in the relationship between the global scale of environmental harm and the structures of racism and colonialism through which that harm and that scale have been historically realized. But rather than pursue this as a question of uneven impacts or responsibility for climate change, I focus on how reading vulnerability in the Anthropocene links it with racism and colonialism. In this section I specifically consider how reading vulnerability operates in step with an expanding dimensionality of power. As the case of the narwhal reminds us, linking vulnerability to new scales of power is an old strategy. It is no accident that an animal Europeans marked with death (the corpse whale) facilitated an imaginative expansion (the unicorn) deployable by European sovereignty in the abstract (the coat of arms, the throne) and on the scale of the geographic (Arctic exploration). Fantasies of erasure enable the material rescaling of power.

In the case of the Anthropocene, the relationship between vulnerability and an expanding dimensionality of power is figured, in two senses, in the negative. First, Chakrabarty describes an agency inaccessible via existing identity categories—"a global approach to politics without the myth of a global identity, . . . a 'negative universal history'" ("Climate," 222). Second, as in Dana Luciano's brilliant "The Inhuman Anthropocene," it is negative in the sense of discerning a new scale of harm: "the Anthropocene is, at base, a political strategy. . . . [I]ts intent is not simply to carve humanity's name upon the stratigraphic map (humans, after all, invented the map in the first place), but to raise awareness of the negative planetary impact of certain human activities, with the intent of altering or mitigating them."[13] While these invocations of negativity suggest different disciplinary and political stakes, both demonstrate how rescaled agency is reconceived through an encounter with the unthinkable, whether historical method or political future.

Negativity in both senses is instrumental to scaling up theories of agency, as is grandiosely metaphorized in Roy Scranton's *Learning to Die in the Anthropocene*. Responding to a familiar set of questions—"We're fucked. The only questions are how soon and how badly," "how we are going to adapt to life in the hot, volatile world we've created"—Scranton, an Iraq war veteran, posits: "we need a new vision of who 'we' are."[14] For him, the conditions for this ontological invention are deathly: "we have to learn to die not as individuals, but as a civilization" (21). The pivot between individual and group death is one Scranton makes around his experience in war. In order to survive he had to convince himself that he "was already dead," a gesture he scales up on the occasion of the Anthropocene: "the greatest challenge . . . [is] understanding that this civilization is already dead" (22, 23).

The concerning ironies of Scranton's proposal are apparent: his own learning to die cannot be shared with those who literally died and continue to die because of U.S. invasion, just as the Indigenous and racialized communities victimized by global U.S. empire will not benefit from Scranton's call for "this civilization" to learn to die in order to endure. More interesting is the arrangement of vulnerability, power, and reading that underlies the metaphor "learning to die" in order to survive—an arrangement Mbembe's theory of necropolitics can clarify.

The spatial terms of Mbembe's argument are crucial to my thinking about vulnerability and new dimensionality of power in the Anthropocene. Broadly, "necropolitics" describes the array of ontological operations through which the sovereign subject is produced in and through death. This includes the way subjects are historically rearticulated through death: "the human being truly *becomes a subject* . . . in the struggle and the work through which he or she confronts death. . . . Becoming subject therefore supposes upholding the work of death. . . . Politics is therefore death that lives a human life."[15] It also includes the attribution of the deathly to living humans figured outside sovereign subjectivity. This explains the coproduction of sovereignty, subject, and racism in broad terms: "the function of racism is to regulate the distribution of death and to make possible the murderous functions of the state" ("Necropolitics," 17). And in the context of slavery specifically, where the racialized nonsubject is defined as embodied labor, "the slave is therefore kept alive but in a *state of injury* in a phantom-like world of horrors and intense cruelty and profanity. . . . Slave life . . . is a form of death-in-life" (21).

Mbembe shows that this process unfolds through the reproduction of the subject via expanding time and capital and through the spatial expansion of colonialism: "Space was therefore the raw material of sovereignty and the violence it carried with it. Sovereignty meant occupation, and occupation meant relegating the colonized into a third zone between subject and objecthood" ("Necropolitics," 26). Space, here, is not just a passive theater for the performance of necropolitics. Necropolitics rearranges and redefines space around power, which Mbembe calls "writing on the ground a new set of social and spatial relations" (25).[16] Mbembe turns to Frantz Fanon to demonstrate how the spatialization of necropolitics generates mappable coordinates of power: "Fanon describes . . . a division of space into compartments . . . the setting of boundaries and internal frontiers epitomized by barracks and police stations. . . . But more important, it is the very way in which necropower operates: 'The town belonging to the colonized people . . . is a place of ill fame, peopled by men of evil repute. They are born there . . . they die there, it matters not where, nor how. It is a world without spaciousness'" (27). For Mbembe, Fanon's necropolitical space is a mappable arrangement ("a division of space into compartments") of bodies and power that can be read ("the town . . . is a place of ill fame") and abstracted ("it matters not where"). It is also a way of reading such that vulnerability becomes symbolic, and spatial meaning describes a certain kind of life, "a world without spaciousness." I argue that it is in these terms that we are given to read vulnerability in the Anthropocene; despite a new scale of harm or potential harm, our epistemic encounter with who, where, and what vulnerability is, is necropolitical. As a consequence, we forfeit the chance to theorize the epistemic value of vulnerability differently and to reexamine the politics of our reading practices.

"A world without spaciousness," for many theories of the Anthropocene, is the conceptual hinge between located experiences of vulnerability and a global enervation of human experience. For Chakrabarty, this is symbolized by the title of Mike Davis's *Planet of Slums* (2006), in which "slums" signifies the scalable inscription of a human future without "freedom" ("Climate," 212). Likewise for Scranton, spatial experiences of harm—"drought and hurricanes, refugees and border guards, war for oil, gas, and food"—scale down to a spaciousless subject: "to live in that world is horrific. . . . The experience of being human narrows to a cutting edge" (*Learning*, 82). In both cases I am interested in how ideas of the world—understood to be free, spacious, or settled—and vulnerability are constrained and made mappable through the idea of the Anthropocene. Reading vulnerability is the idea that disciplinary protocols of apprehension and description naturalize and preserve this relationship.

The question of marking the Anthropocene with a global stratotype section and point (GSSP, a geologic golden spike) is an example of this reading practice. As Stephen Walsh, Felix Gradstein, and Jim Ogg note, GSSPs were adopted as a normalized stratigraphic method in response to H. D. Hedberg's appeals to the International Geological Congress.[17] In discussions about the Anthropocene, GSSPs have been used to debate the year at which we understand the Anthropocene to have begun. Crutzen and Stoermer initially suggested 1784 (the Industrial Revolution) as a starting point,[18] but more recent proposals have shifted it: 1610, 1945, 1952. As Charles Hepworth Holland observes, the origin of this specific metaphor, the "golden spike" that would be driven into and thereby inscribe a given geologic layer, is not clear. However, he notes that the metaphor is widely and plausibly understood to have been borrowed from the ceremonial conjoining of the Union Pacific and Central Pacific railroads in 1869 by Leland Stanford—a gesture in the colonial history of the United States that might be said to have joined the symbolic expression and infrastructural regulation of the geographic expanse of the United States.[19]

As the golden spike was for Stanford, the question of the Anthropocene's GSSP expresses a new dimensionality of human power—humankind as a geologic force. As a way of reading intensifying experiences of environmental harm, the GSSP debate remaps vulnerability on a geologic scale, making spaciouslessness the framework through which that harm is interpreted. This gesture fossilizes responsibility and inaugurates newly constrained concepts of vulnerability and subjectivity (the species). Beyond this, debates about the Anthropocene's GSSP reduce our reading practices to necropolitical terms: predisposing our interpretive practices to reinvest in the dynamics of expanding power through geographically and racially distributed dying. In an era of unavoidably intensifying environmental harm, I want to be able to read more and read better than this scaling up allows.

**Reading Vulnerably**

Robert Warrior's *Tribal Secrets* asks "critics [to] find . . . a way of making ourselves vulnerable to the wide variety of pain, joy, oppression, celebration, and spiritual power of contemporary American Indian community existence."[20] I have suggested that reading vulnerability is problematically ascriptive—a way of arranging subjectivity, space, and power around a scalable construct of the vulnerable. It positions reading as an administrative practice that reduces what vulnerability could be, what it could mean, and what politics it could inspire. These dynamics extend into the space of reading itself, where it does not make sense to ask already differently vulnerable readers to intensify their vulnerability through any single or coordinated reading process. Instead: how could we read Warrior's "making ourselves vulnerable" so that it did not reconstrain agency to those forms of subjectivity—the possessive individual (our selves), the human (ourselves)—through which global environmental harm is executed, distributed, and worried about?[21] Is there a way we could read Warrior's construction, on the occasion of the idea of the Anthropocene, to question those interpretive structures that derive power by writing themselves around vulnerability?

If, for instance, we take "vulnerable" as an adverb modifying "making ourselves," then "ourselves" could no longer be read as a stable or scalable position affected by "the wide variety of pain, joy, oppression, celebration, and spiritual power." Instead, "making ourselves" would indicate an unfinished action directed toward or attentive to an array of meaning-making processes ("pain, joy," and so on). To read Warrior this way would be to understand "vulnerable" differently—as a positive epistemic attitude toward difference.

This approach runs counter to Chakrabarty's call for a "negative universal history" by rethinking the stakes of the insufficiency of subject formations occasioned by intensifying environmental harm. "Making ourselves vulnerable to" is encountering "pain, joy, oppression, celebration, and spiritual power" without reducing their meaning to how they would constrain or advance the interest of a subject. Instead this invites us to revisit the scene of our "making ourselves" out of experiences that forms of subjectivity in crisis were always meant to manage, distribute, and secure against. In the context of this chapter, I want to be careful not to use the idea of reading vulnerably to return to a critique of the subject, which has been done more skillfully and intensively by others.[22] Rather, I use this as an opportunity to develop and deepen a positive epistemic attitude between ourselves-in-the-making and -unmaking and those experiences of living and thinking through environmental harm that cannot be mapped onto or around the subject on any scale.

I return now to okpik's volume with two caveats. First, I read her work as expressing, not evincing, the argument about reading vulnerably I just summarized. *Corpse Whale* already works as a theory of vulnerability, agency, and reading—which I hope only to recenter within conversations about the Anthropocene. Second, my analysis does not come close to capturing the nuances of that argument as expressed in the fullness of okpik's book. What follows can only be suggestive.

The structure of *Corpse Whale* is seasonal. The volume begins:

Raven in the midnight sun
*Siqinq*: Sun January
*Siqinyasaq tatqiq*: Moon of the coming
Sun (3)

and proceeds through monthly/seasonal sections to a final section that starts:

*Ukiuk*: Winter *Siqinrilaq Tatqiq*: December
Moon with no sun. (87)

okpik's use of seasonality as a structure distinguishes her work from poetic forms (the serial, the sequence) that are extricable from the environment. At the same time, seasonality is not thematic in *Corpse Whale*; its poems are not pastoral or antipastoral. These are not eclogues.

The seventh section begins,

*Inyukuksaivik Tatqiq*: July
Moon when birds raise their young
plump breasted snow geese swaddle papier
mâché eggs. Her/my mother's feather blanket
nestles an albino seal. Years ago, it could
have been her/me nestled safely in a snow cloth.
She/I wait/s for the next ten thousand years of
fossil replica, she/I wait/s until my adoption
has worn off her/my mind where pixels turn into
satellites and nightmares.

Season does not indicate a span of time or even a stable temporal pattern in this poem. Rather, it is shaped by at least seven temporalities that are woven through each other: (1) "*Inyukuksaivik Tatqiq*," (2) "July," (3) "Moon," (4) memory: "Years ago, it could / have been her/me," (5) geology: "ten thousand years of / fossil replica," (6) colonial biopolitics: "my adoption," and (7) recovery: "has worn off." Similarly, what place means gathers in tangles of discourse and images: the almost ethnographic "Moon when . . . snow geese swaddle" resolves into symbol ("papier / mâché eggs") only to shift again into "seal," and then "her/me" "nestled" in "feather blanket" or "snow cloth." All of this ultimately disperses again back through ground ("fossil replica"), space ("satellites"), and psyche ("nightmares"). And throughout, the poem's subject is scored by a forward slash: not alienated but bound together across spans of time, space, species. Season, then, does more than divide the volume into twelve sections. It frames a theory of agency and environment that the poems in each section push to its limit.

Reading across seasons/sections of *Corpse Whale*, the dispersal of agency across time, space, and subject gathers momentum. It is no accident that okpik targets these constructs. As I suggest in "Reading Vulnerability" above, these are key sites that link the ascription of Indigenous affective experiences to a scalable concept of death. Mishuana Goeman argues that controlling the meaning of time, space, and subject is essential to colonial power in general, what she calls the "settler grammar of place." She writes: "The classification of 'Indian' has everything to do with spatial occupation of land and bodies," a project that entails "constraining people in places and in bodies that are marked and unmarked in ways that make them legible or illegible."[23] "A settler-colonial grammar," she argues, "prefer[s] to have Indians in a ubiquitous space, controlled by problematic imagining" ("Disrupting," 239). Reading vulnerability in the Anthropocene might be redescribed in Goeman's terms as a settler colonial grammar of environment.

Many poems in *Corpse Whale* explicitly address the link between legible and scalable environmental measures and Indigenous death, which I called earlier the colonial necropolitics of vulnerability. okpik's poem "Stereoscope," for instance, allegorizes the Victorian technology of the stereoscope as an apparatus designed to administer the terms of epistemic encounter:

Stereoscope:

        A device by which two
photographs
        of the same object takes
slightly
        different angles are viewed
together
        giving an impression of depth
slightly wide
        *as in ordinary human vision.* (33)

The poet asks what a system of epistemic encounter that both mimics and replaces "ordinary human vision" is designed to produce. okpik suggests that the technological reproduction of "depth" collapses the complexity of intersubjective experience—this is a rare moment in *Corpse Whale* when pronouns are disaggregated. "She/I" is compressed into an "I" under the examination of a "you": "you sing and I bleed out slowly" (33). The poem continues:

> How is it you conjure          seabird specimens
> from my skull
> . . . . . . . . . .
>           How do you mention
> one by one           our place-names
> and dates of birth  by labeling      your glass boxes?
> How do you print on heavy flat paper
> the artifact of *Nuiqsut*     which makes me
> pull apart from my cartilage? (34)

The poem parodies and subverts the synthetic reading of the stereoscope, pulling a viewable "I," "our," and "me" apart from the viewing "you" by linking the projects of scientific taxonomy to subject formation—an argument powerfully summarized:

>       You see  she/I think/s
> you would      like to extend the ghost surface
>
> of this row    of death loops (34)

I read this passage as a critique of the scalability of colonial reading practices normalized as objective assessments of place, subject, temporality. "The ghost surface" is a world made readable by reducing what Goeman calls the "multiple scales with various intersecting story lines" ("Disrupting," 240), which animate Indigenous environments, to manageable forms of space, time, and subject—okpik's "death loops." This observation resonates with Tsing's critique of scale: "scalability requires that project elements be oblivious to the indeterminacies of encounter; that's how they allow smooth expansion. Thus too scalability banishes meaningful diversity, that is diversity that might change things" (*Mushroom*, 38). As Tsing makes clear, the stakes are more than the erasure of particular instances of life or thought. Scalability is a problem if it diminishes our ability to conceive of how to "change things." This idea aligns with my broader concerns about the relationship between reading and the Anthropocene.

Reading vulnerably asks if it is possible to leverage a moment of critique like this one into reading practices that intensify our encounter with life and thought described as pathological or vulnerable: reading practices that are nonscalable. This is also a central interest of Goeman, who observes: "While many have discussed the 'ruins of representation' and the forceful role they play in the everyday life of Native people, there is a lack of attention to the constitutive construction of body and space" ("Disrupting," 237). Goeman's precise pivot from the "ruins of representation" to the "constitutive construction of body and space" indicates the importance of centering Indigenous life and thought as analytics in an era of intensifying environmental change. Reading vulnerably is an incitement to be socially and intellectually reconstituted in the coconstruction of body and space beyond colonial representations of power and exposure.

In closing, I want to point to two dimensions of the "constitutive construction of body and space" that unfold across sections/seasons of okpik's volume. The first is that dynamic constructions of body and space do not constitute a world separate from that mapped as vulnerable: they rub against it; through friction, body and space are reinvented in the motion of critique. As Tsing writes, nonscalable projects do "not occur in some time before scalability. [They are] dependent on scalability—in ruins . . . [such projects] depend on scalability and its undoing" (*Mushroom*, 40).

A simple example of this might be to take seriously the idea that a work of innovative poetry like *Corpse Whale* should be centered as a theoretical text key to thinking about the Anthropocene—a suggestion that stretches environmental studies' interest in interdisciplinarity to its limit. However, if we are persuaded that there is a necropolitical operation to projects that read vulnerability, then a poem like "A Cigarette Among the Dead" (part of the December season/section) proves theoretically generative. In this poem okpik reconsiders the relationship among vulnerability, agency, and death during the long darknesses of an Arctic winter:

> And from        the macabre marring        and mania which hems
>     her/me in        and cleaves    her/my wretched blue        heart she/I
> stare/s        into the night.
>
> Dazed by the lantern        she/I detach/es from        my carcass.
>     She/I indulge/s in a cigarette        among the dead (90)

This encounter is jarring and melancholy, not manageable: nothing less than a feeling of existential threat. This feeling the poem intensifies ("she/I detach/es from my carcass") and then reframes as a communion ("a cigarette among the dead"). The threat of death as loss of agency is never denied here; it is an inescapable problem of reading, which the subject's body trembles to confront:

>                                                             my hand
>
> shakes as    she/I reach/es the ashtray    to flick. (90)

This communion reveals or relieves nothing ("which eludes her/my eyes swirling") but still prompts the subject's pivot out, not into a scaled-up subjectivity, but into the life existing (beyond view) in vibrant darkness:

> The air breathes                         and delves into    my bloodless
>
> bones each       decomposing, cast down        in the dark
> cast down                in the light                              to gladden
> the shades. (90)

Read as a theoretical reconsideration of colonial necropolitics, the poem reactivates vulnerability as a site of ontological and social motion: not a mappable space of spaciouslessness but, in Goeman's words, a spatially and bodily constituting that is "partial and never complete" ("Disrupting," 242).

The final dimension of the "constitutive construction of body and space" that I want to draw out concerns the epistemic value of the sense of intimacy expressed on the occasion of vulnerability, evident throughout the collection. A poem ominously titled "The Fate of Inupiaq-like Kingfisher" (part of the April season/section) dramatizes this. Of okpik's poems, this is one of the most formally spare and syntactically simple. It begins with the image of a "bird spear" (both animal and weapon suggested together) striking an animal, "piercing depilated skin." Its force amplifies as the symbolic complexity of the "bird spear" ("made of notched bone, / feathered arrows pinnate / around the shaft") meets with a body utterly exposed, "piercing depilated skin." This is a moment of intimacy both in what it describes—an act of fishing, whaling—and as a performed conveyance of meaning. Assuming the subject of this hunt is a narwhal, the poem blurs "whale" between noun and verb, animal and practice, archive and interpretation.

In *The Intimacies of Four Continents* Lisa Lowe retheorizes intimacy beyond the mappable/locatable condition of liberal interiority.[24] Intimacy is an analytic, for Lowe, that exposes the narrative complexities of political and social histories in the aftermath of the global-scale expansion projects of colonialism, slavery, and immigration undertaken by and on behalf of the liberal subject. For Lowe, intimacy refers to the trajectories of encounter, vulnerability, and collaboration that exceed the capture of colonialist epistemic and political forms, but that are no less subject to management under the rubric of liberal intimacy as a privileged sign of interiority and freedom. Intimacy is the friction between the nonscalable and the ruins of scalability. In okpik's poem, interspecies and transhistorical motion is intimate:

Some humans weave themselves

with lime grass,
into large orbs.

Others make goosefeet baskets

of seaweed or with narrow leaves,

or collect matches or tobacco.
. . . . . . . . . . . . .
my existence becoming a flicker

like the orange scales of a kingfisher.

We pirouette, diving, diving,
    deep. (27)

This poem traces moments attentive to choreographies of bodily and spatial constitution that are unstable and unfinished. These are banal and enduring gestures that are also what living in an era of global environmental vulnerability—across political, environmental, and human histories—is like. These intimate shifts across subjectivity, species, and scale do not demand bigger theories of ourselves but rather invite us to consider how reading can become an occasion to reweave ourselves toward what Warrior calls "the wide variety of pain, joy, oppression, celebration, and spiritual power," which colonial forms of subjectivity and history have never been able to constrain.

## Notes

1. Barry Lopez, *Arctic Dreams* (1986; New York: Vintage, 2001), 128 (hereafter cited parenthetically in text).
2. Qtd. in Odell Shepard, *The Lore of the Unicorn* (New York: Random House, 1988), 261.
3. Qtd. in Todd McLeish, *Narwhals: Arctic Whales in a Melting World* (Seattle: University of Washington Press, 2014), 11.
4. L. Weilgart, "A Review of the Impacts of Seismic Airgun Surveys on Marine Life," submitted to the CBD Expert Workshop on Underwater Noise and Its Impacts on Marine and Coastal Biodiversity, London, February 25–27, 2014, https://www.cbd.int/doc/meetings/mar/mcbem-2014-01/other/mcbem-2014-01-submission-seismic-airgun-en.pdf.
5. Stephen Pax Leonard, "The Disappearing World of the Last of the Arctic Hunters," *Guardian*, October 2, 2010, http://www.theguardian.com/world/2010/oct/03/last-of-the-arctic-hunters.
6. d. g. nanouk okpik, *Corpse Whale* (Tucson: University of Arizona Press, 2012), 65 (hereafter cited parenthetically in text).
7. An unduly biographical reading might understand okpik's "she/I" as thematizing her adoption away from her Inuit family. It has been well documented that adopting children out of tribal communities is a colonial tactic that disrupts Indigenous epistemic and social continuities. However, okpik's split pronouns also stretch subjectivity across species, kinship, and geography. Thus this gesture cannot only signify fracture or loss, if for no other reason than that these dispersals are generative: they cue knowing and acting that would otherwise remain obscured.
8. Paul J. Crutzen, "Geology of Mankind," *Nature* 415.23 (2002): 23.
9. Dipesh Chakrabarty, "The Climate of History: Four Theses," *Critical Inquiry* 35.2 (2009): 197–222, 206, 218 (hereafter cited parenthetically in text).
10. Simon L. Lewis and Mark A. Maslin, "Defining the Anthropocene," *Nature* 519 (March 2015): 171–80.
11. Judith Butler, *Notes Toward a Performative Theory of Assembly* (Cambridge: Harvard University Press, 2015), 139.
12. Anna Lowenhaupt Tsing, *The Mushroom at the End of the World: On the Possibility of Life in Capitalist Ruins* (Princeton: Princeton University Press, 2015), 38 (hereafter cited parenthetically in text).
13. Dana Luciano, "The Inhuman Anthropocene," *Avidly: Los Angeles Review of Books*, March 22, 2015, http://www.avidly.lareviewofbooks.org/2015/03/22/the-inhuman-anthropocene.
14. Roy Scranton, *Learning to Die in the Anthropocene: Reflections on the End of a Civilization* (San Francisco: City Lights, 2015), 16–17, 19 (hereafter cited parenthetically in text).
15. Achille Mbembe, "Necropolitics," *Public Culture* 15 (2003): 11–40, 14–15 (hereafter cited parenthetically in text).
16. The interrelationship of these operations is far from as straightforward as slavery being deployed to capitalize land cleared of Indigenous people (as it is sometimes framed). Eve Tuck, Allison Guess, and Hannah Sultan usefully show how the relationship among settler colonialism, slavery, and land is subtle and dynamic in "Not Nowhere: Collaborating on Selfsame Land," *Decolonization*, June 26, 2014, https://decolonization.wordpress.com/2014/06/26/not-nowhere-collaborating-on-selfsame-land.
17. Stephen Walsh, Felix Gradstein, and Jim Ogg, "History, Philosophy, and Application of the Global Stratotype Section and Point (GSSP)," *Lethaia* 37.2 (2004): 201–18, 202.
18. Paul Crutzen and Eugene Stoermer, "The 'Anthropocene,'" *Global Change Newsletter* 41 (May 2000): 17–18.
19. Charles Hepworth Holland, "Does the Golden Spike Still Glitter?," *Journal of the Geological Society* 143 (1986): 3–21, 212.
20. Robert Warrior, *Tribal Secrets: Recovering American Indian Intellectual Traditions* (Minneapolis: University of Minnesota Press, 1995), 114 (hereafter cited parenthetically in text).

21. Here I refer to theories of subjectivity that overinvest in the autonomous individual, for example, C. B. Macpherson's "possessive individualism" in *The Political Theory of Possessive Individualism: Hobbes to Locke* (Oxford: Oxford University Press, 2010), and those (e.g., posthumanism) that seem to wish away the epistemic and ontological differences of race, indigeneity, gender, ability, and so on.

22. Scholars of black and colonialism studies show how the liberal subject is constructed through histories of empire and slavery and manages what human experience is, what it can mean and enact. See Saidiya Hartman, *Scenes of Subjection: Terror, Slavery, and Self-Making in Nineteenth Century America* (Oxford: Oxford University Press, 1997); Denise Ferreira da Silva, *Toward a Global Idea of Race* (Minneapolis: University of Minnesota Press, 1997); and Alexander Weheliye, *Habeas Viscus: Racializing Assemblages, Biopolitics, and Black Feminist Theories of the Human* (Durham: Duke University Press, 2015).

23. Mishuana Goeman, "Disrupting a Settler-Colonial Grammar of Place: The Visual Memoir of Hulleah Tsinhnahjinnie," in *Theorizing Native Studies*, ed. Audra Simpson and Andrea Smith (Durham: Duke University Press, 2014), 236 (hereafter cited parenthetically in text).

24. Lisa Lowe, *The Intimacies of Four Continents* (Durham: Duke University Press, 2015).

# 11.

# Accelerated Reading
## Fossil Fuels, Infowhelm, and Archival Life

*Derek Woods*

As a period of literary history, the decades following the Second World War align with the Great Acceleration, an unprecedented spike in human impact on the Earth system.[1] One striking feature of this period is its overwhelming quantity of reading material, whether in the form of books or other media. Driven by fossil fuels, the scale of the archive has increased by several orders of magnitude since preindustrial times. As Rob Nixon notes, the Anthropocene's "high-speed planetary modification" comes with mutations of cognition brought about by "a digital world that threatens to 'infowhelm' us into a state of perpetual distraction."[2] Under these media-historical conditions, even readers with carefully delimited specializations face a growing shadow archive that we might call the "Great Unread." For those who realize that they want to read more than they can read, the Great Unread provokes a feeling of desire mixed with loss.

Earth's biodiversity is often compared to a library or an archive. However, as Ursula Heise suggests, asking how many species exist during the current mass extinction is comparable to "asking how many books are archived in a library on fire."[3] No matter how one quantifies biodiversity—whether the unit is species, genomic variation, symbiotic assemblage, or morphological difference—writers on extinction constantly remind us that "we" are burning the archive faster than biologists can catalog it and much faster than they can interpret each book in detail.[4] Yet as Heise notes, the relationship between biodiversity and archival life is one of inverse proportion: "even though we live in an age of mass extinction, the number of species is steadily increasing—the number of known species, that is" (*Imagining Extinction*, 56).[5] For Dominic Pettman, there is thus a "troubling correspondence between an explosion of animal images at the very moment of a mass implosion of actual species."⁶ The diversity of "life forms" (a term preferable to "species" in this context and one that takes on a specific meaning below) paradoxically increases as it decreases because there are more biological entities legible as names, images, descriptions, and traces in the archive of stored memory as the Anthropocene goes forward. The dodo has become a cliché reproduced in thousands of texts *because* it went extinct. Despite the fact that a *decline* in biodiversity accompanies the growth of the archive of life forms, the emotional affect it creates is perhaps similar to that experienced by many readers facing the Great Unread. The structure of feeling that makes this affect public is one of desire provoked by the multiplicity of life forms, but desire cut with a sense of loss which registers the growth rate of the Great Unread and the countervailing rate at which human activities homogenize biodiversity.

This chapter describes a reading method suited to this relation between extinction and infowhelm, a way of conceptualizing and interpreting the footprint of the biosphere in the archive. Franco Moretti's theory of genre and "morphospace," a concept borrowed from evolutionary theory, suggests an approach to composing the archive of life forms represented in literary history. The morphospace of archival life is made up not of whole texts, but rather of fragments which ultimately compose an Earth-scale meaning system shot through with cultural and linguistic disjunctions. In the planetary framework of the Anthropocene, this archive of life forms demands a Vernadskian reading method. Vladimir Vernadsky was the Russian scientist who first understood the biosphere in a systemic sense, as a totality of living beings producing the same environment to which they adapt. As Will Steffen and his co-authors note, Vernadsky's biosphere and noosphere concepts are aspects of the conceptual genealogy of the Anthropocene ("Anthropocene," 844–45). While no single reader is able to scan the full archive of life forms, to make it present as meaning, a Vernadskian reading approaches texts as if they are part of an Earth-scale entity coupled to the biosphere. In this context, the force of the Kantian "as if" is that we can think this Earth-scale archive, but not access it directly or in its totality—a point that makes this Vernadskian reading *partial* and distinguishes it from holistic accounts of the Earth system.

### Infowhelm as Petro Reading

There are many ways to understand the who and what of the Earth system mutations that the term "Anthropocene" seeks to name (there is no need to recite the growing list of terms suffixed by *-cene*). As Tobias Menely and Jesse Oak Taylor argue in the introduction to this collection, critics and Earth system scientists have developed countervailing geological and ecological interpretations of the Anthropocene concept.[7] There is, for example, an emerging opposition between interpretations that stress the human stratigraphic signal and those that stress our *systemic* influence on the biosphere. Paleontologist Mark Williams and his co-authors give substance to the latter interpretation when they argue that the "Anthropocene biosphere" is the product of four factors: (1) a widespread "resetting of ecosystem composition and structure," (2) human appropriation of the biosphere's energy productivity and the combustion of fossil fuels, (3) the evolution of nonhuman species through artificial selection, transgenics, and the alteration or creation of habitat, and (4) increased coupling of the biosphere with the rapidly evolving "technosphere."[8]

As other contributors in this book argue, there is much to be said for a plurality of Anthropocene signals. Nonetheless, there is broad consensus that the period following the Second World War shows major quantifiable changes in the Earth system, and the Anthropocene Working Group's recommendation of a mid-twentieth-century global stratotype section and point (GSSP) identifies the Great Acceleration as the threshold of the epoch.[9] Such arguments suggest that this historical discontinuity shapes many cultural phenomena in the period. Accordingly, and expanding on Kate Marshall's suggestion that we consider contemporary literature in terms of "geological forms of periodization,"[10] I understand the contemporary Anthropocene biosphere as a historical frame in which we see a new relation among biodiversity and the life forms inscribed in a fast-growing archive. Since fossil fuels are a necessary (if not sufficient) condition for the Anthropocene biosphere, they are the final element in the structure of feeling central to this chapter.[11]

One of the "footprints" or "imprints" that humans leave on the biosphere (and the nonhuman semiosphere) is neither $CO_2$ nor concrete but communication, whether stored, in process among psychic systems, or created in acts of reading. As population, energy use, and the distribution of media technologies spiked following the Second World War, so has the quantity of "human" communication, simplifying and replacing nonhuman semiotic systems as it has grown.[12] In 2010, engineers working for Google Books estimated the number of published titles on Earth (not editions or copies) at 129,864,880.[13] Comparing this number to the quantity of books circa 1700 is difficult and perhaps, as book historians tell me, impossible due to incomplete records. Yet we can infer from archives such as Early English Books Online, which contains most of the titles (125,000) published in England by 1700, that the *global* curve is steep (if not uniform) between 1700 and 2010. Taking the world's largest library as another exemplary microcosm, the annual reports of the librarian of Congress show growth in the book and print collections from about 100,000 items in 1866 to 1 million in 1900, 15 million in 1966, and 38 million in 2014.[14] From these data, the planetary number of books grew by more than two orders of magnitude between the early Industrial Revolution and the present. If we add other media to the equation, the curve would only be steeper, and the code read by programs and computers would add extra orders of magnitude.[15] Despite the absence of precise numbers for the quantity of books and other media, it is clear that the graphs of literary output, communication in every medium, and the archive of nonhuman life would all display a similar shape to those that chart human energy use and the rate of anthropogenic species extinction.

In the Great Acceleration, reading takes place in the context of an exponential increase in the quantity of human and machine communication. For many readers, the feeling that accompanies this abundance is "infowhelm," first defined in the *Oxford English Dictionary* in 2009 as "apathy, indifference, or mental exhaustion arising from exposure to too much information." Yet this use of "information" is imprecise in light of Claude Shannon and Warren Weaver's distinction between information as a statistical quantity and meaning as the semantic effect that arises for readers of signs.[16] For them, information and meaning are not aligned: a document containing "dog" repeated over and over could contain the same number of bytes as Haraway's "Cyborg Manifesto." *Information* is not what "exhausts" us, because "exposure" to information turns it into meaning. As media theorist Bruce Clarke puts it, "the trace is not self-observing," but there is a tendency, in our theoretical present, "to off-load the duties of cognitive systems onto information structures," ascribing agency to texts and forgetting Shannon and Weaver's distinction.[17] There is no meaning without readers. These readers need not be human, but they must be cognitive systems of some kind because "meaning results from coupled networks of informatic and cognitive processes" (Clarke, *Neocybernetics*, 25).[18] The difference between information and meaning matters because it situates reading's limits within a new epistemic frame. In this frame, signification is only loosely coupled with the statistical and physical properties of information.

The Great Unread affects how it feels for people to read even what they *can* read, with implications for the practices and fantasies of reading that emerge in response to infowhelm (or, more accurately, *semiowhelm*). One such fantasy of uninhibited data transfer into human brains is A. E. van Vogt's *The Voyage of the Space Beagle* (1950), the story of an interstellar research vessel composed only of STEM departments. The only exception to this rule is Elliott Grosvenor, a "nexialist" who joins the mission when his discipline is still new and misunderstood. His job as nexialist is to synthesize knowledge from all the disparate sciences represented by members of the expedition. Grosvenor's field has no proper content, only a powerful method for learning and synthesis that he describes as "applied whole-ism."

He learns through a series of unconscious download techniques: prepared by "hypnosis and psychotherapy" to "break [his] initial resistance," he uses sleep machines to fill his memory with knowledge.[19] Other methods include "little films where each picture stays on but a fraction of a second" (*Voyage*, 55), like the subliminal messaging techniques made famous by Vance Packard's *The Hidden Persuaders* (1957). *Ex machina*, a fantasy of unimpeded download replaces what would otherwise be a welcome method for reading across fields, circumventing the relative slowness of reading and interpretation, which transpose information into meaning. With these enhancements, however, the nexialist synthesizes the knowledge of his shipmates, becoming powerful enough to defeat a series of terrifying life forms and the chemistry department.

In light of the sheer quantity of information and potential meaning facing readers, Grosvenor is an example of what Thierry Bardini calls *Homo nexus*: a symptomatic figure in the Great Acceleration and a response to the feeling of loss provoked by the unread.[20] No doubt the roots of this feeling go as far back as the early years of what Marshall McLuhan calls the "Gutenberg galaxy," leading to figures such as the "Renaissance man"—a term coined by Victorians that referred retroactively, as professionalization became dominant during the late nineteenth century, to a period with less specialization and fewer books.[21] Yet the mid-twentieth century multiplied cognate figures and institutions: Van Vogt's *Homo nexus* draws from the philosopher Alfred North Whitehead (for whom "nexus" is a central concept); the semiotician Alfred Korzybski's theory of general semantics; first-wave cybernetics (with its military-industrial synthesis of disciplines); ecosystem ecology; and other midcentury efforts to overcome social complexity and specialization—efforts that persisted in late twentieth-century developments from David Foster Wallace's *Infinite Jest* (1996) to InfoVis, Earth system science, and the Intergovernmental Panel on Climate Change.[22] While the problem of having too much to read is long-standing, this dilemma becomes acute in the decades since the Second World War, when the episteme of "information" leaves few fields of knowledge untouched.[23]

The exponentially increasing archive to which *Homo nexus* responds is brought to you by coal and oil. Without fossil fuels and the media technology they enable, the archive and the rate of communication could not grow as they have.[24] More than simply "extensions of man" (McLuhan), media technologies, as Jussi Parikka argues, are "extensions of the earth," dependent on industrial labor for minerals, fossil energy, and petrochemicals.[25] *Homo nexus* is thus a petro reader with close ties to the energetic conditions for the production of information, an accelerated reader able to overcome the time restraints of embodied subjects. As the potential energy of fossil fuels unravels, the speed of history increases. This means that the Great Acceleration is far "longer" (or, at any rate, larger) than any other period of literary history. Fossil energy makes possible more history, more communication, and more inscription per unit time than in the past.

In *The Voyage of the Space Beagle*, this situation appears in the nexialist's response to his geologist colleague's skeptical curiosity about nexialism. Referring to the sleep technique, the geologist says that learning "all that is known" would take "just under a thousand years," suggesting that Grosvenor should admit to this constraint on generalist reading (54). The geologist's millennium expresses the notion that there is *more time* as history accelerates, making the new durations embedded in the archive unmanageable for individual readers. Grosvenor presents the fantasy of recompressing, like a zipped file, this thousand-year portion of the tempest (or "time storm," as Jeffrey Jerome Cohen puts it elsewhere in this volume) released from fossil fuels. But as the geologist learns more about the speed of nexialist cognition—namely, that it allows a rate of learning closer to "an entire science at one sitting"—his "pursed lips" and "staggered expression" are the affects of a slow learner overwhelmed by transhumanist reading practices (55). As a character written at the outset of the Great Acceleration, the nexialist is a compensatory response to the feeling of loss that many readers experience in relation to the period's exponentially growing quantity of information—to which the only antidote appears to be circumventing reading altogether.

**Reading the Third Morphospace**

Combined with its focal character, van Vogt's title, *The Voyage of the Space Beagle*, evokes our structure of feeling that links fossil fuels, extinction, and infowhelm during the Great Acceleration. As an allusion to Darwin's *The Voyage of the Beagle* (1839), the title connotes the study of biodiversity along with the new geographical distances and awareness of extinction that, for Darwin, make concrete species spatially and temporally contingent. By adding "space" to Darwin's title, van Vogt repeats the long-standing analogy between space travel and exploration/colonization by sea. Yet as a work of science fiction, *The Voyage of the Space Beagle* also transposes the biosphere into literary history, imagining alien life forms by analogy with extant and extinct species of Earth. The novel thus fits Stefan Helmreich and Sophia Roosth's argument that science fiction contributes to the disarticulation of "life" and "form" during the history of "life form" as a keyword. In combination with fields such as theoretical morphology, transgenics, astrobiology, and synthetic biology, science fiction participates in a conjectural rationality concerned with possible forms of life. In these genres and fields, abductive reasoning (in Charles Sanders Peirce's sense of reasoning from as-yet-unrealized premises) combines with "forms as yet unencountered" and "notions of life materializing in physical spaces of possibility," such as planets radically different from Earth.[26] In order to understand the relation of life, form, and the actual species composing Earth's biodiversity, we need a concept for the interaction between the biosphere and what Yuri Lotman calls the semiosphere—that is, for the footprint of the biosphere in the archive.[27]

For David Raup, an evolutionary biologist, the concept of morphospace yields a thought experiment and a model of possible variations in shape within a group of species. As philosophers James Maclaurin and Kim Sterelny put it, models such as Raup's assume that "we can represent morphology as a multidimensional space, with each dimension of that space corresponding to a variable morphological feature" (*What Is Biodiversity?*, 59). Raup developed his model using basic parameters to create a geometrical space that contains both the actual shell shapes of mollusks and possible shapes that have never been observed. He used an enormous 1960s computer to vary the shell parameters, creating "the museum of all shells."[28] In theory, such "hyperspaces" have a very high dimensionality, since each parameter has its own dimension in the model. In practice, theoretical morphologists have only become adept at modeling possible forms using a small number of parameters from a small number of taxonomic groups. As Maclaurin and Sterelny argue, there is no conceivable way to model the "space of all possible morphologies" (79). Partial morphospaces, centered on taxa as they exist in a particular historical moment, are the only computable ones—despite our ability to think the global morphospace which they enticingly imply.

In *Graphs, Maps, Trees*, Franco Moretti adapts the concept of morphospace for the "evolution" of literary genres, adding a second, cultural morphospace by analogy with Raup's model. Moretti attempts to explain how "the micro-level of stylistic mutations," such as the clue, have led to new genres, such as the detective novel.[29] "Explanation" is a privileged term in his approach, which seeks a scientific criticism distinct from hermeneutic methods.

Moretti situates these stylistic mutations in a Darwinian tree which represents their variation and selection. In his application of morphospace to literary history, he points to the "market" as an (undefined) agent of selection. Following the classical anthropologist Alfred Kroeber, Moretti argues that the branches of the tree of life only diverge, while the branches of the "tree of knowledge" also permit fusion (*Graphs*, 79).[30] In this way, "divergence prepares the ground for convergence" in the tree of culture. Yet as multitudes of these variations and selections occur simultaneously, they cross a difference of scale: "this system of differences at the microscopic level adds up to something that is much *larger* than any individual text: . . . the genre—or the tree—of detective fiction" (76). With this claim, Moretti reasserts his argument for distance: a bird's-eye view of the relation between history and form. The forces that shape literary history are "the very small and the very large," "devices and genres," not whole books or the authorial oeuvre (76). Distant reading *composes* the second morphospace out of small-scale formal variations with the objective of explaining large-scale trends in literary history.

Following Raup and Moretti, I propose a third morphospace to think the relation between extinction and archival life during the Great Acceleration. The third morphospace includes every instance of biological life represented in literary history and in the broader audiovisual archive, from Shakespeare's "midnight mushrooms" to Thomas Hardy's heather and Octavia Butler's octopus-like aliens. While Raup's model visualizes possible and actual shell shapes, in the third morphospace the fantastical replaces the possible. Monsters, deities, and alien life forms, so often imagined as mutations or as combinations of human and nonhuman species, vary around types derived from terrestrial biomes. "Morphospace" is a good term for this archive because it is so often the shapes and colors of nonhuman species—their morphologies, not their DNA—that leave a mark in cultural memory. As an Earth-scale phenomenon and a region of the semiosphere, the third morphospace is the sum of all actual and imaginary life forms in communication, whether they operate thematically, rhetorically, or structurally.[31]

If the first morphospace is natural and the second cultural, then the third (in an argument characteristic of much Anthropocene discourse) corresponds to neither of these categories. While many theorists no longer want to refer to this third category as a hybrid of nature and culture, the basic stakes are clear: as the "human" semiosphere increases in size, it mixes more thoroughly with the nonhuman systems outside of it, paradoxically including and excluding them as it comes into contact with them. The question then becomes to what extent morphologies alter and constrain the possibilities of meaning as humans communicate about them, so that, like animal domestication, semiosis is more than a one-way process controlled by human beings.[32] In this reading, the third morphospace is a more apt transposition of Raup's concept from biology to culture than Moretti's. The former is limited the possibilities of naming, describing, and rhetorically manipulating life form and so avoids straining the analogy between biological morphology and literary form.

Taking a Vernadskian approach to the third morphospace situates it at the confluence of the biosphere and the semiosphere. One workable method for studying it is a form of partial reading suggested by modernist collage. This method decontextualizes examples of archival life form, searching for patterns not legible in individual works. As opposed to Moretti's distant reading, partial reading skims under constraint in order to find or compose previously unknown archives within the total archive, which no reader (including Google) can read as a whole. Drawing on Moretti, the method builds up from "microscopic" systems of difference—here, fragments of morphospace—to larger-scale systems that appear only when the signed, titled literary works they inhabit fall away.

In this short chapter, my morphospace is only a snippet view. The source texts are novels and poems published during the Great Acceleration, between the Second World War and the present, in Canada, Britain, and the United States. Beyond these constraints, which reflect my research area, I use no conscious or consistent search criteria:

> 1. His great *fore-legs* twitched with a shuddering movement that arched every razor-sharp *claw*. The thick *tentacles* that grew from his shoulders undulated taughtly. He twisted his great *cat head* from side to side, while the *hairlike tendrils* that formed each ear vibrated frantically, testing every vagrant breeze, every throb in the ether.

There was no response. He felt no swift tingling along his intricate *nervous system*. There was no suggestion anywhere of the presence of the id creatures, his only source of food on this desolate planet. Hopelessly, Coeurl crouched, an enormous *cat-like figure* silhouetted against the dim, reddish sky line, like a distorted etching of a *black tiger* in a shadow world. (Van Vogt, *Voyage*, 5, emphases added)

2. But in lambda 117 the inhabitants' evolutionary descent had been from a *radial form*, something like a *five-armed starfish*, itself developing out of a *spiral*. They were in several sizes from a few centimetres to almost [a] metre across. . . . Among their main artefacts were wide but low buildings, profusely decorated on the undersides of their roofs, mostly with *spirally fungal and rooting forms*. . . .

It is only in circumstances like these when we realize how much we ourselves are constructed *bi-laterally* on either-or principles. *Fish rather than echinoderms.* I know this in a sense, but when it came to attempted communication it increased my difficulties far more than I ever thought it would.[33]

3. The site is a charred patch becoming green with *new weeds*, inside a *copse of beech and some alder*. Camouflaged metal stands silent across a crowd of late *dandelions, gray heads* nodding together waiting for the luminous wind that will break them toward the sea, over to Denmark, out to all points of the Zone. Everything's been stripped. The vehicles are back to the hollow design envelopes of their earliest specs, though there's still a faint odor of petrol and grease. *Forget-me-nots* are growing blue violet yellow among the snarl of cables and hoses. *Swallows* have built a *nest* inside the control car, and a *spider* has begun filling in the *web* of the Meillerwagen boom with her own.[34]

4. *Buzzworm* had a thing for *palm trees* too. Maybe 'cause they're skinny and tall. Maybe if he put his ear to the trunk of a *palm tree*, he could hear the radio waves descending from the *scraggly fronds* at the top. And he really knew his *palm trees*. *Family Palmaceae. Four thousand species.* Tall ones called *Washingtonia Robusta* or *Mexican Fan Palm.* Similar ones with thicker trunks were called *Washingtonia Filifera.* Mostly you noticed the tallest *Robustas. Buzzworm* was always talking about them like he was their personal gardener. You caught him staring at *palm trees*, seemed like he was talking to them.[35]

5. The universe is expanding. I cannot go back to *labyrinthodonts. Tooth-writing* my sediments. Shale, sandstone, yanked thin by mountains drifting west, grinding under tectonic plates. I cannot go back to *Silurian conodonts, Paleocene creodonts, the*

*tree-toothed, the bright-toothed, the nail-toothed, the humped, crested-, breasted-, the mill-toothed mylodons, the leaf-toothed sharks, the squalid-toothed whales, the mixotoxodons, the hypsilphilodons, the loxolophodons, the ungulate-rodent-carnivore tillodonts, the socket-toothed thecodonts, or the tooth-writing ants* who ringed their gates with *molars from thousands of tiny mammals*.[36]

Different from orderly lists of taxa or representational descriptions, this small region of the third morphospace consists of rhetorical operations on anatomical terms and the names of organisms, extant or extinct. "Cat-like" and "something like a five-armed starfish" (1, 2) combine simile and anamorphosis to describe imaginary extraterrestrial beings. Starting with a simple image, simile stretches the cat or the starfish into a distortion of itself. The two fragments begin with a well-known organism and transform it in a way that invites readers to envision a constrained possibility space: not an earthly cat, but not a lizard or a formless being. Examples of crasis such as "leaf-toothed sharks," the "gray heads" of "dandelions," and "ungulate-rodent-carnivore tillodonts" (5, 4, 5) use one life form metaphorically to characterize another. Yet the catalogs of palm trees and extinct mammals (4, 5) create a synecdochic effect. In context, "Washingtonia Robusta" and "loxolophodon" fan out toward a broader diversity of forms. Synecdoche is a major trope within the structure of feeling described above, since it causes readers to imagine an unwritten archive of morphological variation beyond the forms that are actually named, undermining the presence of the latter while producing desire for the former. Denotative nouns such as "weeds" and "swallows" (3) raise the following question: at what levels of taxonomic abstraction does literature inscribe finer-grained species distinctions as conventional signs? If nowhere near the full diversity of morphologies present in the biosphere registers in the human semiosphere, and far fewer in literary history, then what causes a certain level of abstraction to be legible—and for which readers?

We can study the third morphospace as a subset of literary history to understand how life form splits away from the morphologies of nonhuman species to become a semiotic phenomenon, contingent on the system of differences that produce meanings. For Helmreich and Roosth, drawing on Raymond Williams's keyword approach, the historical semantics of "life-form" and "life form" reflect this process of abstraction.[37] Since the early nineteenth century, these terms have "pointed to the space of possibility within which life might take shape" ("Life Forms," 19). With this emphasis on the distinction between actual and possible, their concept of life form has much in common with Raup's original model of morphospace. Yet the possibility spaces surrounding organisms adumbrate not only future evolutionary trajectories, but the indexical (fossils, tracks) and arbitrary signs associated with organisms, including the thematic and rhetorical meanings actualized by readers. These meanings are constrained, but not determined, by the same actually existing organisms that constrain the possibilities of the *first* morphospace. For example, in the sentence fragment "fish rather than echinoderms" above (2), we can read "fish" and "echinoderm" in opposition only because of a third element, the "ourselves" that refers to our bilaterally symmetrical primate bodies. In this context, "fish rather than echinoderms" can modify "ourselves" by setting up a distinction between animals that have a back and front and radial animals, such as starfish. The form of the human body (but nowhere near the full complexity of it, nor its DNA) limits the contextual meanings of "fish" and "echinoderm." Whether in response to paleontology, astrobiology, theoretical morphology, or science fiction, the result of situating life in the third morphospace (which grows with the archive over the course of the Great Acceleration) is to make life seem contingent and spectral, pointing to alternative possibilities and undermining its presence. Had Derrida written *Specters of Darwin* along with *Specters of Marx* (1994), the famous "doubtful contemporanaeity of the present with itself" would describe how the meaning systems associated with organisms divide present biodiversity between the past and the future and between presence and absence.[38]

**Ecopolitics and Partial Reading**

Throughout *Welcome to the Anthropocene*, a digital animation geared to climate change awareness, we view the rotating Earth from space. A disembodied British voice narrates the *longue durée* of human impact as an accompanying graph curves toward its (increasingly canonical) exponential moment.[39] "People swarm to cities" as we arrive at the Great Acceleration, which appears as an unprecedented spike in the graph. This biopolitical use of the social insect swarm to characterize the human collective is a symptom of the view from above sustained throughout the animation, as the surface of the Earth fills with electric lights and the flight paths of a million jets.

Stacy Alaimo notes the prevalence of the view from above in Anthropocene discourse, calling instead for situated perspectives on contemporary geological, atmospheric, and biological mutations.[40] For her, the view from above reinscribes what Donna Haraway critiques as the "god-trick" of the view "from nowhere": giving a single, holistic account of an epoch better understood as differently read and felt by embodied subjects, human and nonhuman.[41] Eschewing a focus on the geological and terrestrial effects of the Anthropocene, Alaimo attends to the chemical, biological, and aquatic dimensions of the Anthropocene biosphere. Like Haraway's privileging of partial perspective in "Situated Knowledges," partial reading is a method uninterested in distance. The partial reader's ability to think the third morphospace as an Earth-scale entity, while reading only small parts of it, is different from pursuing definitive explanations of what Moretti calls "the collective system as a whole" (*Graphs*, 4).

The third morphospace is not an ark. Arks are synecdochic: they stand for the biosphere's full spectrum of biodiversity and complement the incomplete project of classification inaugurated by colonial natural history. The association of Moretti's "more rational literary history" with distant reading is the central problem with his return to structuralism and the notion of a scientific literary criticism, which avoids engagement with the complication and politicization (but not abandonment) of objectivity undertaken in science studies since the 1980s (*Graphs*, 4).

One aspect of this work has been the critique of colonial science as what Elizabeth DeLoughrey and George B. Handley call "an empirical and imperial project" of taxonomy and collection "constituted by the flora, fauna, and human knowledges extracted from the colonies."[42] As Helmreich notes, the function of the view from above is often to own and control territory (*Sounding the Limits*, 131). The sense of loss that paradoxically accompanies the growth of the third morphospace is a function not only of infowhelm and ongoing extinction, but of the historical complicity of the tools that record it colonial violence.

Although the method outlined above relies on "collecting" and decontextualizing archival life form, it resists the notion that Anthropocene reading should construct a total or even representative view of biodiversity's semiotic footprint. The point of this method is not to create synecdochic menageries nor to respond to the limits of reading with nexialist data management. The point is to pose new questions about how life form functions in literary history, how the broader archive (de)couples with the biosphere, and what role archival life might play in ecopolitics. In this way, the method shares something in common with Juliana Chow's "critical partiality" in this volume, which favors dispersal over the holism that has always been a factor in ecological thought, and with Bruce Clarke's second-order account of an Earth system that no longer resonates with the term "whole Earth" common in Anthropocene discourse.[43] My version of partial reading is Vernadskian—despite the fact that Vernadsky wrote the book on Earth-scale holism—because it starts from the assumption that one *is* reading part of an Earth-scale entity. I can read a zone of the third morphospace under this assumption without expecting to arrive at the "collective system as a whole." Thus, the method pries the Earth scale away from its synonymy with the "whole Earth" conceived via the view from above.

Alaimo's chapter "Your Shell on Acid" suggests one way which the third morphospace becomes political (*Exposed*, 143–68). Her discussion concerns the implications of ocean acidification, an effect of climate change that dissolves the shells of mollusks and kills coral reefs. Her reading of Nina Bednaršek's micrographs of pteropod shells, one intact and one in the process of dissolving, helps me return to Raup's "museum of all shells" in an Anthropocene frame. In this way, the particular reflexivity at work in the this frame demands that we see "ourselves," however divided this subject might be, as a biogeochemical agent. For Williams and his co-authors, one driver of the Anthropocene biosphere is precisely this extension of "artificial" selection to every corner of the Earth, intentional or (more often) otherwise ("Anthropocene Biosphere," 207). The arrow that at first points *from* the biosphere *to* the semiosphere loops back to constrain the becoming of the biosphere, as the third morphospace affects the evolutionary trajectory of the first. That is, if humans influence the biosphere in the ways that Earth system scientists now describe, and if cultural production shapes human behaviors, laws, and geopolitical systems, then practices such as partial reading may themselves shape the Earth's future biodiversity, however indirect this link between reading and morphological change might be. Indeed, if modelers such as Raup were to embrace the Anthropocene frame, they would need to factor the third morphospace into the possibility spaces they construct, regardless of its status as artifice.

The third morphospace has grown during the sixth mass extinction. There are more species *for us* (for some of us, or for the archive) as nonhumans vanish. With that in mind, perhaps there are better and worse ways to read archival life form as fossil fuels run out. Reading literary history for its work with life form is a composition practice as much as a method for studying a vast preexisting archive. If communication depends to some degree on the operation of the biosphere's traces in the semiosphere, it must be possible to make this dependence stronger, building long-term, transgenerational attention to life form in ways that make the natural history museum and the BBC's *Planet Earth* look like curious morality plays. And if literary traditions depend, to any degree, on the semantic affordances of morphology, then we do not know what possibilities of meaning are latent in Earth's biodiversity in a time when many still take the obsolete animal-vegetable-mineral ontology as fact. Partial to fungi, the composer Václav Hálek points in the right direction by "recording" the songs of hundreds of mushroom species. So does the care for mammals, birds, and reptiles in Philip K. Dick's *Do Androids Dream of Electric Sheep?* (1968), which is all but absent from director Ridley Scott's cyberpunk adaptation of the novel in *Blade Runner* (1982). Over the coming millennia of climate change, there could be centuries-long poetic traditions devoted to Ascomycota, competing avant-gardes who reject algae for beetles, and multispecies collages to replace the twentieth-century legacy of junkspace.[44] Such practices would not invite an untenable biocentrism that claims to treat all life equally, but a political aesthetic that inevitably privileges some life forms over others.

## Notes

1. According to one group of Earth scientists, "although the imprint of human activity on the global environment was, by the mid-twentieth century, clearly discernible beyond the pattern of Holocene variability in several important ways, the rate at which that imprint was growing increased sharply at midcentury." Will Steffen et al., "The Anthropocene: Conceptual and Historical Perspectives," *Philosophical Transactions of the Royal Society A* 369 (2011): 842–67, 849 (hereafter cited parenthetically in text).

2. Rob Nixon, *Slow Violence and the Environmentalism of the Poor* (Cambridge: Harvard University Press, 2011), 12.

3. Ursula Heise, *Imagining Extinction: The Cultural Meanings of Endangered Species* (Chicago: University of Chicago Press, 2016), 27 (hereafter cited parenthetically in text). See also Daniel C. Dennett's Borgesian "Library of Mendel," in his *Darwin's Dangerous Idea: Evolution and the Meanings of Life* (New York: Simon and Schuster, 1995), 107–13.

4. James Maclaurin and Kim Sterelny discuss the ambiguities that attend efforts to quantify biodiversity in *What Is Biodiversity?* (Chicago: University of Chicago Press, 2008), 1–26 (hereafter cited parenthetically in text). Ursula Heise notes the same problem in *Imagining Extinction* (25–26). A related problem is the difference between population and species diversity. Huge population losses could be devastating for humans and nonhumans alike without resulting in species extinction.

5. For twenty-first-century data on extinction rates, see Rodolfo Dirzo et al., "Defaunation in the Anthropocene," *Science* 345.6195 (2014): 401–6; and Richard Monasterky, "Life: A Status Report," *Nature* 516 (2014): 159–61. For an uncommon counter to the sixth-extinction narrative, see Stewart Brand, "We Are Not Edging Up to a Mass

Extinction," *Aeon*, April 21, 2015, https://aeon.co/essays/we-are-not-edging-up-to-a-mass-extinction.

6. Dominic Pettman makes this point in a gloss of John Berger's *Why Look at Animals?* (2009) in Pettman, *Human Error: Species-Being and Media Machines* (Minneapolis: University of Minnesota Press, 2011), 244.

7. Will Steffen and his co-authors work to unite these approaches in "Stratigraphic and Earth System Approaches to Defining the Anthropocene," *Earth's Future* 4 (2016): 324–45.

8. Mark Williams et al., "The Anthropocene Biosphere," *Anthropocene Review* 2.3 (2015): 196–219 (hereafter cited parenthetically in text). See also Menely and Taylor's discussion of Clive Hamilton's work in the introduction to this volume.

9. See Colin Waters et al., "The Anthropocene Is Functionally and Stratigraphically Different from the Holocene," *Science* 351.6269 (2016): 137–52. Will Steffen and his co-authors reinforce this position, but update their previous (Steffen et al., "The Anthropocene") graphs to account for the differential "consumption" in "wealthy countries," "those with emerging economies," and "the rest of the world." As they write, "Only beyond the mid-20th century is there clear evidence for fundamental shifts in the state and functioning of the Earth System that are beyond the range of variability of the Holocene and driven by human activities." Steffen et al., "The Trajectory of the Anthropocene: The Great Acceleration," *Anthropocene Review* 2.1 (2015): 81–98, 81. By contrast, Donna Haraway cites developmental biologist Scott Gilbert in arguing that the "Anthropocene" is better considered a "boundary event," like the K-Pg boundary, than a proper geological epoch. Haraway, *Staying with the Trouble: Making Kin in the Chthulucene* (Durham: Duke University Press, 2016), 206 (hereafter cited parenthetically in text).

10. Kate Marshall, "What Are the Novels of the Anthropocene? American Fiction in Geological Time," *American Literary History* 27.3 (2015): 523–38, 529.

11. This relation between oil and the mediation of vanishing life forms is similar to that noted by Stephanie LeMenager in Upton Sinclair's novel *Oil!* (1927), in which "aesthetic images and environmental emotions that valorize driving and even the process of oil extraction" become "modes of facilitating the body's capacity for self-extension toward other life." LeMenager, *Living Oil: Petroleum Culture in the American Century* (New York: Oxford University Press, 2014),

8. For a supporting formulation about oil and biodiversity (which are metonymically the "same" thing), see Elizabeth Kolbert, *The Sixth Extinction* (New York: Holt, 2014), 2.

12. My scare quotes register posthumanist theories of communication, such as that of Cary Wolfe, who suggests that we "reconceive language in terms of the dynamics of *différance* that, because they are fundamentally inhuman in both their technicity and their extension to extrahuman processes of communication, institute the inhuman at the human's very origin." Wolfe, *Animal Rites: American Culture, the Discourse of Species, and Posthumanist Theory* (Chicago: University of Chicago Press, 2003), 74.

13. Leonid Taycher, "Books of the World, Stand Up and Be Counted! All 129,864,880 of You," *Google Books Blog*, August 5, 2010, http://booksearch.blogspot.com/2010/08/books-of-world-stand-up-and-be-counted.html.

14. "The Annual Reports of the Librarian of Congress," https://catalog.hathitrust.org/Record/000072049.

15. In 2010, for example, Walmart's 2.5-petabyte database held 167 times the amount of data in the Library of Congress. "Data, Data Everywhere," *Economist*, February 25, 2010, http://www.economist.com/node/15557443.

16. Claude E. Shannon and Warren Weaver, *The Mathematical Theory of Communication* (Urbana: University of Illinois Press, 1949), 99–100.

17. Bruce Clarke, *Neocybernetics and Narrative* (Minneapolis: University of Minnesota Press, 2014), 25 (hereafter cited parenthetically in text).

18. See Bernhard Siegert's account of the transition from "antihermeneutic" to "postherneneutic" media theory in Germany:

*Cultural Techniques: Grids, Filters, Doors, and Other Articulations of the Real*, trans. Geoffrey Winthrop-Young (New York: Fordham University Press, 2015), 6–17.

19. A. E. van Vogt, *The Voyage of the Space Beagle* (New York: Manor, 1950), 54–55 (hereafter cited parenthetically in text).

20. Thierry Bardini, *Junkware* (Minneapolis: University of Minnesota Press, 2011), 146 (hereafter cited parenthetically in text).

21. Marshall McLuhan, *The Gutenberg Galaxy* (Toronto: University of Toronto Press, 2011). For early modern infowhelm, see Elizabeth L. Eisenstein, *The Printing Press as an Agent of Change* (Cambridge: Cambridge University Press, 1979).

22. I take the references to Whitehead and Korzybski from Bardini (*Junkware*, 146). Heather Houser discusses David Foster Wallace as an author of infowhelm in "Managing Information and Materiality in *Infinite Jest* and Running the Numbers," *American Literary History* 26.4 (2014): 742–64.

23. For John Johnston, "information proliferation" is a technological force that incites a formal response in the postwar novel. *Information Multiplicity: American Fiction in the Age of Media Saturation* (Baltimore: Johns Hopkins University Press, 1998), 1–8. See also Paul Stephens, *The Poetics of Information Overload: From Gertrude Stein to Conceptual Writing* (Minneapolis: University of Minnesota Press, 2015).

24. Niklas Luhmann's alternative to this Marxist picture of infrastructure driving the increase of communication would attribute some of this growth to the recursive self-production of communication itself. See, for example, Luhmann, *Theory of Society*, trans. Rhodes Barrett (Stanford: Stanford University Press, 2012), 1:113–250.

25. Jussi Parikka, *A Geology of Media* (Minneapolis: University of Minnesota Press, 2015), 17–25.

26. Stefan Helmreich and Sophia Roosth, "Life Forms: A Keyword Entry," in Helmreich, *Sounding the Limits of Life: Essays in the Anthropology of Biology and Beyond* (Princeton: Princeton University Press, 2016), 19–35, 32–33 (hereafter cited parenthetically in text).

27. A Soviet semiotician, Lotman reasoned by analogy with Vladimir Vernadsky's indispensable biosphere concept. For Lotman, "all semiotic space may be regarded as a unified mechanism" for which "primacy does not lie in one or another sign, but in the 'greater system' . . . outside of which semiosis itself cannot exist." Lotman, "On the Semiosphere," *Sign System Studies* 33.1 (2005): 205–29, 208. My thanks to Adam Webb-Orenstein for this paper.

28. David Raup, "Computer as Aid in Describing Form in Gastropod Shells," *Science* 138 (1962): 150–52.

29. Franco Moretti, *Graphs, Maps, Trees* (London: Verso, 2005), 91 (hereafter cited parenthetically in text). Moretti derives the concept from Stephen Jay Gould, not from Raup.

30. This is a distinction that Lynn Margulis and others have explicitly critiqued since the beginning of research into "symbiogenesis," or evolution by symbiosis, in the late nineteenth century. See Lynn Margulis and Dorion Sagan, *Acquiring Genomes: A Theory of the Origins of Species* (New York: Basic, 2002). Emily Apter puts it differently in *Against World Literature* (New York: Verso, 2013): "does differentiation (in the species sense) necessarily come at the expense of hybridity models of cultural difference?" (50).

31. Douglas Kahn uses the term "earth magnitude" to describe the ionosphere's electromagnetic "whistlers" and the artistic and scientific practices of humans intrigued by them in *Earth Sound, Earth Signal: Energies and Earth Magnitudes in the Arts* (Berkeley: University of California Press, 2013), 16.

32. For a brief account of animal and plant domestication as a form of symbiosis, see Alfred Crosby, *Children of the Sun: A History of Humanity's Unappeasable Appetite for Energy* (New York: Norton, 2006), 25–34. Anthropomorphism as pathetic fallacy is one way of reading the third morphospace as a one-way process.

33. Naomi Mitchison, *Memoirs of a Spacewoman* (Glasgow: Kennedy and Boyd, 1962), 11, emphases added.

34. Thomas Pynchon, *Gravity's Rainbow* (New York: Penguin, 1973), 560, emphases added.

35. Karen Tei Yamashita, *Tropic of Orange* (Minneapolis: Coffee House Press, 1997), 30, emphases added.

36. Meredith Quartermain, *Nightmarker* (Edmonton: NeWest, 2008), 35, emphases added.

37. Raymond Williams, *Keywords* (New York: Oxford University Press, 1985).

38. Jacques Derrida, *Specters of Marx: The State of the Debt, the Work of Mourning, and the New International*, trans. Peggy Kamuf (New York: Routledge, 1994), 39. See also Timothy Morton, *The Ecological Thought* (Cambridge: Harvard University Press, 2010), 61–62.

39. Owen Gaffney and Félix Pharand-Deschênes, dirs., *Welcome to the Anthropocene* (2012), https://www.youtube.com/watch?v=fvgG-pxlobk, which has more than two hundred thousand views on YouTube. The animation echoes the *Economist*'s May 2011 leader of the same title.

40. Stacy Alaimo, *Exposed: Environmental Politics and Pleasures in Posthuman Times* (Minneapolis: University of Minnesota Press, 2016) (hereafter cited parenthetically in text).

41. Donna Haraway, "Situated Knowledges: The Science Question in Feminism and the Privilege of Partial Perspective," *Feminist Studies* 14.3 (1988): 575–99, 589.

42. Elizabeth DeLoughrey and George B. Handley, eds., *Postcolonial Ecologies: Literatures of the Environment* (New York: Oxford University Press, 2011), 11. See also Richard Grove, *Green Imperialism: Colonial Expansion, Tropical Island Edens, and the Origins of Environmentalism, 1600–1860* (Cambridge: Cambridge University Press, 1995); and Mary Louise Pratt, *Imperial Eyes: Travel Writing and Transculturation* (New York: Routledge, 1992).

43. Bruce Clarke, "Rethinking Gaia: Stengers, Latour, Margulis," *Theory, Culture, and Society* (January 17, 2017), http://journals.sagepub.com/doi/abs/10.1177/0263276416686844. In Hamilton, for example, "Earth System science is the science of the whole Earth as a complex system beyond the sum of its parts." Clive Hamilton, "Getting the Anthropocene So Wrong," *Anthropocene Review* 2.2 (2015): 102–7, 102. One thing that distinguishes second-order from first-order systems theory is the former's critique of holism and organicism.

44. Donna Haraway discusses a beautiful example in *Staying with the Trouble* (76–81): the Crochet Coral Reef of the Institute for Figuring in Los Angeles (http://crochetcoralreef.org).

## 12.
# Climate Change and the Struggle for Genre

*Stephanie LeMenager*

The search for Anthropocene genres has been energetic, contentious, and popular. People outside of academia, people who might not be expected to care about genre, are looking hard for Anthropocene genres—for patterns of expectation and narrative form with which to combat this unsettling era of climate shift and social injury. Much that we know—and I will risk this "we" to indicate humans broadly—about life and its parameters falls apart in the shadow of global climate change. Habit, the subjective practice of reality, frays in this unique moment of global ecology, and such fraying indicates a potential shift in human understandings of the everyday. Ultimately in this chapter I turn to the novel as one genre of the everyday, but I begin with a humbler genre, the news. The news and the novel have been paired before, as generators of imagined community (namely, the nation). In most cases, nations can no longer be said to coalesce around either genre in an era of screen culture and self-selecting publics. For those whose reading is restricted to sources chosen to bolster an already decided vision of the real, the news will not "happen," as it does for those of us who still enjoy the surprise of turning a page, either on a screen or in print. To the extent that this chapter anticipates the possibility of surprise, it addresses those who still seek it—and the extension of intersubjectivity that surprise might entail—by reading what they have not curated so closely as to exclude any inconsistencies in ideology or worldview.

For sociable readers, who want to encounter and dwell with others, one look in the newspaper hints at a world unlivable by previous standards: "sunshine flooding" in southern Florida eighty days of the year, the Iranian city of Bandar Mahshahr reaching a heat index of 165 degrees Fahrenheit, the Great Barrier Reef nearly dead.[1] As a genre, the news has long been a dystopian account of the everyday, anecdotes of catastrophe digestible with coffee, a training ground for forgetting all sorts of endings, including mortality. But climate change "news" fails to be "news" insofar as it implies an end to the everyday itself, since the everyday relies on human habit and its complement of forgetting. Extreme weather, including superstorms and severe drought, and all of these conditions that are taking hold *as conditions* rather than as events shift the ground of habit and call attention to the profoundly ecological, interdependent state of humanity. Climate change represents, among other things, an assault on the everyday. This means that it will register in those genres intended to query and explore what is understood as probable, cyclical, and even trivial experience.

The everyday implies getting by, living alongside the world, living through it. One of the reasons that the American protoenvironmentalist Henry David Thoreau hated "the news" was because it implied that "the world" is a disposable externality, a serial fiction with an iterative and forgettable plot. (Thoreau's extensive recording of seasonal variations in his journals registers the apparently small differences within cycles in such a manner as to show profound respect for those nonhuman agents—for example, berrying bushes—that are irreducible to human plotting.) As Stacy Alaimo and other new materialist thinkers remind us, the environment is not an externality. Toxic off-gassing from our carpets, intestinal bacteria, and climate all make clear that the world lives inside of us, and we it.[2] This realization, theorized as transcorporeality (bodies moving across, through, and inside one another) by Alaimo,[3] comes to the doorstep and to ordinary talk in climate change news. In a kind of anti-obituary in the weather section of the *New York Times*, victims of Hurricane Sandy are described as deindividualized bodies arrayed amid storm-fallen trees and flooded streets.[4] This is news without the therapeutic structuring of plot, the obituary denied the familiar arc of an individual life. It is an Anthropocene genre, the weather section obituary. In it, human lives disappear into a map of unprecedented storm damage that can be anticipated as symptomatic of a shifting ground condition, a new everyday.

## Cli-Fi as Novelistic Mode

Faced with an everyday in which the habit of living alongside the world crumbles under pressure from material impingements—superstorms, flooding—we are at the edge of something, perhaps modernity, perhaps even "humanity" depending on how you define it. Since climate change began to show itself in the late twentieth century, varied genres have come into being to speak, self-consciously, of human extinction, both real and philosophical. Climate fiction (cli-fi) is the most popular. It is the focus of this chapter, both because of its notoriety and because of its explicit project of redefining humanism and the humanities.

I will not attempt a typology of cli-fi. Fictions that have been called cli-fi are remarkably diverse, from psychological realism to science fiction to newer genres attempting to stand apart from cli-fi, such as solarpunk. Cli-fi ranges across media, including digital, television, film, short fiction, the novel, and memoir. That said, I recognize cli-fi across genres and media as a symptom of a social need, the need for new patterns of expectation, to paraphrase Lauren Berlant's capacious description of "genre."[5] This call for genre comes at a moment when life as many humans know it is changing at a pace and scale difficult to imagine. My discussion of climate change begins in the struggle for genre, by which I mean the struggle to find new patterns of expectation and new means of living with an unprecedented set of limiting conditions. More explicitly, in this chapter I address the ways in which cli-fi aspires to envision a climate change culture for readers who are in some cases losing their sense of what it means to be human, to generate culture, and to love.

This intimate and collective questioning of what it means to be human—reflecting the ontological insecurity that sociologist Kari Norgaard identifies in lived experiences of climate shift[6]—recommends response in what I describe as the "novelistic mode." In Alastair Fowler's classic account of literary forms, genres are structures that coalesce, rigidify, and open out to experimentation and dissolution, while "mode corresponds to a somewhat more permanent poetic attitude or stance, independent of particular contingent embodiments of it."[7] For Fowler, mode is what is left once a genre decays. Structural innovation ceases within a particular genre, the genre disperses into the fuzzier attitude of mode, and the mode in turn may generate newer genres. But cli-fi is a relatively new structural response to changing social and ecological conditions. If anything, it is just now coalescing. My insistence on it as a novelistic mode reflects my interest in mode as "a method, a way of getting something done," in the words of the science fiction scholar Veronica Hollinger, who sees sci-fi in particular as less a genre than a way of living in the world.[8] Cli-fi, I argue, marks another way of living in the world—a world remade profoundly by climate change. Hence, I practice a version of distant reading, surveying numerous works in this chapter not to create a graph or tree of cli-fi's dispersion and evolution, but rather to attempt to capture the social attitudes and desires, the atmospheric mode, of cli-fi.[9]

The novelistic mode offers a method for making social worlds by modeling individual consciousness in relationship with imaginary but possible worlds. Literary historians like Nancy Armstrong have lamented the privatization of human experience by the European novel, its tendency "to attach psychological motives to what had been the openly political behavior of contesting groups."[10] By contrast, cli-fi, like its older sibling sci-fi, indicts the privatization of human experience even as it participates in novelistic modes of action. That is, it makes the social through the vivid evocation of individuals in relationship with possible worlds. The need for cli-fi is profoundly social, although its realization rarely includes a vibrant evocation of collective, rather than individual, experience. Finally, the novelistic mode of cli-fi is not restricted to any subgenre within what Franco Moretti identifies as the diverse system of novelistic genres:[11] cli-fi can be seen as realism, sci-fi, and even memoir, a close relative of the Euro-American novel from its beginnings in epistolary form.

**Setting: The Everyday Anthropocene**

Cli-fi summons and chafes against a neoliberal feeling-state that I call the "everyday Anthropocene." The genealogy of the everyday Anthropocene as a background setting for cli-fi begins, for me, in the *Parable* novels of Octavia Butler. These novels have been widely regarded as speculative fiction and only recently associated with the term "cli-fi," which was coined around 2007, after Octavia Butler's death. Nonetheless, the *Parable* novels offer a social and historical dimension to cli-fi that I find indispensable to the larger project of making climate change publics savvy enough to imagine both thriving and surviving with global climate change.

Butler describes the realization of the everyday Anthropocene (without naming it such) in *Parable of the Talents* (1998), the second novel in her magisterial, unfinished trilogy that began with *Parable of the Sower* (1993). For Butler, the era of humanity's succumbing to the collateral damages of modernity is identified as the "Pox," an outbreak of socioecological disease. "The Pox was caused by accidentally coinciding climatic, economic, and sociological crises."[12] The "Pox" names a relatively short period of time, lasting roughly from 2015 to 2030, in which the world as it was known in the 1990s, when the two *Parable* novels were written, came undone. "Overall, the Pox has had the effect of an installment-plan World War III," writes the narrator of this section of the novel

(*Parable of the Talents*, 8). He is Bankole, an African American doctor who was born in 1970 and can testify to the fact that the Pox has a long history, beginning before 2015. "I have watched as convenience, profit, and inertia excused greater and more dangerous environmental degradation. I have watched poverty, hunger, and disease become inevitable for more and more people" (8). The cultural work of the *Parable* novels involves reconciling the crises of dystopian story structures with the habits of living on, which underwrite what I call the "everyday." As other readers have noted, Butler indicts neoliberal policies, such as privatization and deregulation, as causes of climate change and of the collective and intimate forms of social suffering that it entails, including widespread refugeeism.[13]

The primary protagonist of both *Parable* novels, Lauren Olamina, founds a religion called Earthseed, which emphasizes constant attention to the changes seemingly imposed on us by the material world. We might "shape" these changes if we refuse to forget our embeddedness in them. Olamina's mantra, "shape the change," implies attending vigilantly to the world that enters and impinges on us.[14] As Sylvia Mayer notes, *Parable of the Sower*, in particular, borrows from the prescriptions for collective liberation evident in the genre of the American slave narrative, where escape from bondage depends on both print literacy and environmental literacy ("Genre," 193). Such preparation might take the form of learning Indigenous uses of medicinal plants, communal gardening, and marksmanship. Habits of close attention are integral to Butler's notion of bringing about the interruption of the future, in Fredric Jameson's terms, by which Jameson means utopian thought that truly breaks from ("interrupts") current hegemonies.[15]

Drawing on long histories of colonialism, slavery, debt peonage, rape, and resource extraction, Butler's *Parable* novels "[criticize] the emerging neoliberal world" *and* allow for the possibility of meaningful resistance to it (Streeby, "Speculative Archives," 34). Shelley Streeby places the *Parable* novels in the category of "critical dystopia," where a background condition is deplored, but still the possibility of new sites of resistance can be imagined (34). Thus, Butler's broad setting is the everyday Anthropocene, by which I intend to evoke neoliberalism as both a feeling-state and its underlying socioecological conditions, characterized by enduring crises that never quite come to a head. The philosopher Teresa Brennan has coined the term "bioderegulation" to describe the effects of impossible but undramatic conditions of labor in the wealthier world within late capitalism, conditions that kill slowly by means of stress, sleep deprivation, anhedonia.[16] The cultural critic Rob Nixon writes in a complementary fashion of "slow violence" as it affects the world's poor. This wearing, structural violence has been carried out, over time, against the poor and the places in which they live through the diverse practices of colonialism—from fossil fuel extraction to territorial occupation to, again, economic globalization.[17] Both bioderegulation and slow violence ask us to think through environmental degradation, war, and even extinction without the irresponsible and self-indulgent excitement attached to narratives of apocalypse.

The everyday Anthropocene offers a setting or space-time for bioderegulation and slow violence, and it names my correction to contemporary epochal discourse that capitalizes on the charisma of crisis. By "everyday Anthropocene" I imply the present tense, lived time of the Anthropocene, and I recommend paying attention to what it means to live, day by day, through climate shift and the economic and sociological injuries that underwrite it. Epochs are not attentive to the wearing away of bodies, their slow depletion. Epochs are time monuments, attaching us—by "us" here I mean those elite humans who identify ourselves with world authorship—to stone, to universal history. As such, the idea of epochs works to organize new modes of forgetting. The Anthropocene, conceived as a geologic epoch, is a coping strategy of sorts, externalizing not the world so much as time, such that we can forget the moment-by-moment loss of the world by naming its passing on a geologic scale authenticated (and externalized) by a golden spike. Whether the spike marks the radioactive traces of the atomic age present in rock strata or the sedimentary evidence of reforestation after the seventeenth-century genocide of Native peoples of the Americas, it is an indicator that our own time can be understood as displaced onto an elsewhere in which narrative significance resides. The elsewhere is the stratigraphic record, not history per se, and it is not exactly continuous with ourselves.

In contrast, the everyday Anthropocene offers a more granular and personal account of near catastrophic change that believes not in new worlds or even new humanisms so much as "in this one-shot life and the body." These words come from the African American writer Ta-Nehisi Coates, and they name what I see as the focus of the novel in the era of climate change.[18] The project of the Anthropocene novel, I argue, is at best a project of paying close attention to what it means to live through climate shift, moment by moment, in individual, fragile bodies. It is at best a project of reinventing the everyday as a means of paying attention and preparing, collectively, a project of staying home and, in a sense distant from settler-colonialist mentalities, *making* home of a broken world.

The philosopher of science and longtime SF (both sci-fi and speculative feminist) practitioner Donna Haraway makes a similar point when she writes of "staying with the trouble" as an antidote to the "abstract futurism" that can be imagined in response to the epochal shift of the Anthropocene. "Staying with the trouble requires learning to be truly present, not as a vanishing pivot between awful or edenic pasts and apocalyptic or salvific futures, but as mortal critters entwined in myriad unfinished configurations of places, times, matters, meanings" (*Staying*, 1). Haraway recognizes Butler's *Parable* novels as complementary to her "own explorations for reseeding our home world," for making refuges for those threatened by displacement and extinction (119). I too see the *Parable* novels as a working model for stories to live by—rather than stories to die by—in a time of climate shift and potential social collapse. This project of the novel within the lived time of the everyday Anthropocene has been more explicitly identified by writers of color and by feminist writers and philosophers than by the elites who have not yet experienced what Coates calls "the plunder of our bodies" (*Between*, 37).

Referring to people who imagine themselves to be white Americans ("Dreamers"), Coates writes in his memoir, *Between the World and Me* (2015), in relation to global climate change: "the Dreamers will have to learn to struggle themselves, to understand that the field for their Dream, the stage where they have painted themselves white, is the deathbed of us all. The Dream is the same habit that endangers the planet, the same habit that sees our bodies stowed away in prisons and ghettoes" (152). Coates did not write a cli-fi novel or even recommend the writing of such novels. However, his book recognizes the climate change era as an extension of a protracted struggle, born of many centuries of colonialist history. His intimate relationship to that struggle and the dialectic between individual and social experiences (with each making over the other, through time) in the memoir places it in the novelistic mode, to a degree. *Between the World and Me* shares some of the assumptions about how the Anthropocene arises not as a geologic epoch so much as lived experience, a culmination of historical tendencies, which we see in Butler's "installment-plan World War III." Similarly, the African American poet Jerry Ward Jr. describes "climate change" as "World War III," which is both a "race war" and a war of survival fought by the human race, in his memoir about life in New Orleans after Hurricane Katrina, *The Katrina Papers* (2008).[19]

These African American versions of the Anthropocene note that the Anthropocene cannot be conceived apart from colonialist practice, including racism, the first act of enslavement (predating African American history, of course), the first rape, the first territorial theft. The authors acknowledge that nature, at least in the form of the human body, has never been an externality. And they imply that what we—meaning all humans—face is not necessarily an end, but profound change, struggle, war, and perhaps a reapportionment of socioeconomic burdens and powers. I make note of the everyday Anthropocene in the context of what might be called an anticolonialist or even Black Anthropocene as a means of giving depth to a central preoccupation of all cli-fi, which is the loss of humanity, as authors variously define it, and often also the loss of "civilization"—with, at times, an explicit call for new humanisms and new cultural forms. Not all cli-fi is as attentive to the possible upending of relations of power as are these black and postcolonial Anthropocene visions, however.

**Learning to Die as Novelistic Practice**

Roy Scranton's memoir, *Learning to Die in the Anthropocene* (2015), makes a call for new genres and new forms of "civilization" consonant with the goals of many practitioners and defenders of cli-fi. Like Coates, Scranton writes about his individual experience as a way to converse with and perhaps shift the political field, and he makes his central theme confronting the threat to the self that lurks implicit in climate change. So, I place him as a practitioner of the novelistic mode of climate change representation. Drawing from his experiences as a soldier in Iraq, as an environmentalist, and as a graduate student in the humanities at Princeton, Scranton writes: "In order for us to adapt to this strange new world, we're going to need more than scientific reports and military policy. We're going to need new ideas. We're going to need new myths and new stories, and a new conceptual understanding of reality, and a new relationship to the deep polyglot traditions of human culture.... We need a new vision of who 'we' are. We need a new humanism—a newly philosophical humanism, undergirded by new attention to the humanities."[20] Here Scranton's call to philosophy and the arts is expansive enough to embrace the work of Octavia Butler and Donna Haraway, if we allow that his "new humanism" might be conceived as posthumanism or multispeciesism. Scranton invites broad participation. Potentially, his call could be seen to embrace even Indigenous writers and thinkers like Thomas King, whose cli-fi novel *The Back of the Turtle* (2014) introduces the possibility of a multispecies (post)humanism based in First Nations cosmology but accessible to settler allies.[21]

Where Scranton departs from what we might see as a transcultural, coalitional project of remaking humanism in the era of climate change is signaled by the apocalyptic gesture of his book's title, "learning to die," and his subtitle, "reflections on the end of a civilization." Scranton explains, referring explicitly to the Western philosophical tradition of liberal humanism: "If, as Montaigne asserted, 'To philosophize is to learn how to die,' then we have entered humanity's most philosophical age, for this is precisely the problem of the Anthropocene. The rub now is that we have to learn to die not as individuals, but as a civilization" (*Learning*, 21). Of course, *who* is learning to die as a civilization is a relevant question. The Citizen Potawatomi philosopher Kyle Whyte makes clear that "our ancestors"—meaning not only Potawatomi people but all Native North Americans—have already endured and lived beyond the end of their world; they are working now on shoring up and conserving what is left.[22] King in *The Back of the Turtle* makes a similar argument, if more fancifully, by staging the return of a prodigal Native son, Gabriel Quinn, to the home river in British Columbia that he inadvertently ruined through the release of a lethal pesticide, which, now banned, moves surreptitiously across the oceans on a barge. The barge itself eventually floats back to the home territory carrying its toxic load. Finally, it is pushed and "sung" away from the home shore by Quinn and a community of survivors.

Both King and Whyte argue, essentially, for collectivism and conservationism as means of living with climate change. They imagine coming together in local communities, often watershed-based, to save what can be saved. Octavia Butler's *Parable* novels strike an uncomfortable balance between this kind of local commitment or "taking care of unexpected country," in Haraway's phrase,[23] and a more typically American and transcendentalist imaginary that includes the possibility of space travel and the settlement of new planets.[24] In contrast, much of the cli-fi of Europe, white America, Britain, and Scandinavia echoes Scranton's central theme: living through climate change means, first, "learning to die" or, in another of Scranton's phrases, "letting go."

I mentioned Ta-Nehisi Coates earlier in part because to me it is remarkable that his memoir on race in America ends with a discussion of climate change, but also because in those pages are, I think, some keys to why the theme of learning to die is so prominent in Euro-American (and, arguably, European) climate fiction. Coates's book begins with a discussion of embodiment—black embodiment, America's obsession with black embodiment, and the necessity of accepting the body as life—all of this in order to live as one chooses in a black body. "In accepting both the chaos of history and the fact of my total end, I was freed to truly consider how I wished to live—specifically, how do I live free in this black body? It is a profound question because America understands itself as God's handiwork, but the black body is the clearest evidence that America is the work of men" (*Between*, 12). Putting aside what is by no means a marginal point about the historical creation of black bodies in the United States, I focus on the more implicit idea here that not having a body—not thinking of one's body as oneself—is a privilege denied to people understood to be black, as well as to some other kinds of people.

The privilege of not thinking of oneself as embodied, as matter overwritten and writing history, is a privilege lost to all humans, including those imagined to be white, in the era of climate change. Climate change presents a radical challenge to ways of living once seen as unencumbered by material constraint—for example, living as white, living as hypermobile in a culture of speed, living as a top consumer in an age of credit. Learning to die, as a theme of cli-fi, is always in part the problem of coming back to oneself and one's defining conditions as problems of recalcitrant matter. "Letting go" is losing any trappings of social transcendence—whether these are understood as whiteness, wealth, heteronormativity, or national belonging. When learning to die enters novelistic practice, it assumes particularity—for example, Scranton's experiences in war—that invites readers' empathy, theory of mind, and, to some extent, identification. The theme of learning to die as played out in novelistic practice also tends to mask the contention built into the proposition that there is a "we," a broad if not universal social category, who must learn to die. I argue that a historical argument would be different: there are people in this world who already have learned to die, and there are people who, faced with anthropogenic climate change, are only just now learning to die.

For cli-fi writers in the Euro-American tradition, learning to die implies letting go of some cherished tenet of humanism. Sometimes this letting go proves productive of new aspirations for humanity and human sociality. In other instances, learning to die means recognizing oneself as a biological entity, challenging traditional humanist ideas of the sovereign self. In the English novelist J. G. Ballard's *The Drowned World* (1962), often cited as one of the earliest examples of cli-fi, entering into climate collapse and embracing a new Europe of flood and tropical heat means, for the protagonist Bodkin, "abandoning the conventional estimates of time in relation to his own physical needs and entering the world of total, neuronic time, where the massive intervals of the geological time-scale calibrated his existence."[25] For Ballard, climate change invites a kind of white flight from social and historical responsibilities into a trippy, "neuronic" existence in line with countercultural fantasies of the 1960s. What I sardonically call "white flight" here marks my response to *The Drowned World*, a novel at ease with a tired discourse of racial primitivism. It may be that the effort of imagining a species mind, so to speak, tends to preclude more nuanced understandings of social history and power. The historian Dipesh Chakrabarty in his influential critique of Anthropocene thought worries about whether social injustices can be addressed when one thinks of oneself as a member of a species rather than a historically specific community. To paraphrase Chakrabarty, once we begin to think of ourselves as a species exercising its brute force on the planet, then it becomes quite difficult to think in more nuanced terms about histories of uneven (human) power, including colonialism and the like.[26] Yet Chakrabarty also argues that the thought experiment of feeling oneself, phenomenologically, as part of a species may be necessary to understanding and addressing global climate change.

To know oneself as embodied in a deep, evolutionary sense may bring about both a profound remembering of ecological enmeshment and an invitation to forget one's specific history or relationship to power. Kim Stanley Robinson's monumental Science in the Capital trilogy (2004–7) offers a nuanced recognition of the responsibilities attendant on biological self-realization in the character Frank, a primary narrator whose fascination with sociobiology leads him to reflect on his social motives from the perspective of species tendencies. This profound self-consciousness, which is also a consciousness of the self as species, does not lead to consistently altruistic or just behavior. Yet Robinson's inclusion of a sociobiological plot in a realist series of novels about the marginalization of science in U.S. politics makes an interesting statement about how understanding climate change calls for a surrender, essentially, of the sovereign self to biological notions of being. Bringing science into the U.S. capital, into policy making at the federal level, might require a philosophical sea change in which we begin to recognize ourselves as primates with specific endowments and limitations.

A more literal take on what it means to die into a recognition of oneself as nonsovereign and embodied is found in Cormac McCarthy's *The Road* (2006),[27] a novel that I see as a latecomer to the postnuclear annihilation subgenre of speculative fiction but that has been embraced by environmentalists, including the well-known author George Monbiot, as cli-fi.[28] In *The Road*, the nameless protagonist recounts his descent into embodiment through near starvation and a wearing sickness. For McCarthy, loss of the sovereign self when that self is, as in the case of his protagonist in *The Road*, a father, also implies the potential loss of patriarchal order, morality, and the capacity for love—at least until a new father/god figure can be found. The profound fear of social failure at the core of U.S. masculinity is on display in McCarthy's work, provoked in this novel by a devastating and sudden climatic change, perhaps caused by nuclear detonation, which has rendered the world sterile. Unfortunately, to my mind, McCarthy resolves the crisis of American masculinity and all of the systems that this crisis generated historically (the nuclear family, a Protestantism centered on an all-powerful father, weapons of mass destruction) by simply offering another, healthier father figure—walking out of the dust like a miracle—at the novel's end. Thus, *The Road* ultimately refuses Scranton's invitation, the invitation arguably posed to white America and wealthy America by climate change, to learn to die.

Scranton's prescription for learning to die in the Anthropocene may begin with learning to die as an individual—something he understands intimately through his service in war. But the larger implications of the prescription involve learning to lose one's "civilization," by which Scranton seems to mean both the best that has been thought and said, and the zombie-like habits of consumerism, social media addiction, and political spectatorship. "Learning to die as an individual means letting go of our predispositions and fear," Scranton writes. "Learning to die as a civilization means letting go of this particular way of life and its ideas of identity, freedom, success, and progress" (*Learning*, 24). Given my initial approach to the topic of learning to die in climate fiction, it may not be necessary to ask, again, *whose* civilization must be let go? Nonetheless, I reiterate the question so as not to let it creep too far to the margins of this argument, because cli-fi and discussions of it too often assume an all-inclusive "we." For now, "we" is the European American subject, comfortable enough in wealth, contemplating not only the loss of self-sovereignty but also the end of a kind of culture that has exceeded its ecological carrying capacity.

Two of the most stylistically impressive cli-fi novels to date, the Canadian author Emily St. John Mandel's *Station Eleven* (2014) and the British author Marcel Theroux's *Far North* (2009), enumerate cultural losses in the process of imagining deeply detailed and (by virtue of their imaginative richness) to some extent compensatory futures. Mandel's "incomplete list of things disappeared after the end of the world" includes "cities," "films," "screens shining in half-light," and with the loss of the cultures of print, celluloid, and screen, "no more reading and commenting on the lives of others, and in so doing, feeling slightly less alone in the room. No more avatars."[29] Blazoned with the phrase "Because survival is insufficient," the horse-drawn buses carrying Mandel's protagonists crisscross the Great Lakes region, a relatively safe territory. The protagonists are musicians and actors known collectively as the Traveling Symphony. They perform concerts and Shakespeare plays for those who are left in the former United States of America, the survivors of the Georgia flu. They are, in essence, "staying with the trouble" by practicing those elements of their former civilization that they find still useful—and the useful things include literature and art. The Georgia flu was a pandemic apparently accompanying climate change, and it wiped out most of the human population, ending the modern world. Mandel invests the material artifacts of informational and narrative culture (screens with their eerie light, paper books colored with bright inks) with sensuous power and resonance—in the sense of social memory. The loss of these things does not serve as a scolding referendum on modern consumerism. As Mandel's Museum of Civilization suggests toward the novel's end, the objects that made up the cultures of the wealthier world—the world that accelerated global climate change—hold the keys to redemption as well as the symptoms of shortsightedness and civilizational collapse. Mandel's subtle conceptual response to near extinction inheres in her insistence on not simply letting go of the culture that brought about the "end." The protagonists in *Station Eleven* act as bricoleurs, repurposing what they recognize as still usable from the old culture.

For Mandel, as for Marcel Theroux in *Far North*, the airplane more than any other Anthropocene object signifies the gorgeousness, globalism (affirmative *and* imperialist), and profound arrogance of modernity. "The end of air travel" (*Station Eleven*, 35) is the name Mandel gives to the transitional phase between civilizational collapse and her straggling "afterward." Theroux's protagonist, a woman named Makepeace who passes as a man through most of *Far North*, sets out on the journey that sparks the novel's plot in search of a downed airplane. Another airplane—and the desire to travel on it—brings her to the novel's climactic conflict. Reflecting on her decision to leave her relatively safe compound in search of the downed plane, Makepeace confides: "Seeing that plane the first time in the lake, I'd never known hope like it.... The plane was a sail, luffing and snapping to a new course as it came to find me. I would walk on its warm deck with my pretty feet. There was silk and cloves in its hold, coconuts, oranges. Well, I guess it brought on the hooey in me."[30]

For both Mandel and Theroux, airplanes offer a false promise of human connection and even closeness to "the world," when the world is abstractly conceived in terms of mappable distance and totemic commodities, like oranges. More authentic connections in their post-apocalyptic story worlds are violent, raw, and simply awkward, like the knife fights remembered on the bodies of Mandel's protagonists, or the forgettable sex between Theroux's Makepeace and a fellow wanderer, which produces an unexpected child. Perhaps the point is that we "moderns" have never transcended the profound frictions of social encounter, have never been cosmopolitan or particularly socially intelligent—even in the age of airplanes. Certainly, Ian McEwan's realist cli-fi novel *Solar* (2010) serves as an indictment of human intelligence as it is conceived by wealthy moderns. McEwan offers up the worldly man of science as a philandering, cowardly glutton, a figure who performs the profound vulnerability of the rationality attributed to science.

Nathaniel Rich's cli-fi novel *Odds Against Tomorrow* (2013), which became a sensation in part because of Rich's apparent prediction of Hurricane Sandy through a powerfully imagined New York City flood, investigates the degree to which speculation, in the form of risk assessment scenarios for insurance purposes, is a largely irrelevant form of intelligence too. While Rich's explicit target is the actuarial logic of modern risk society, his send-up of a risk analyst who invents fabulous scenes of infrastructural collapse to some extent indicts the project of the cli-fi novel itself. In *Odds Against Tomorrow*, the combination of imagination and archival analysis that leads to actuarial prediction (and to speculative fiction) cannot compete with geological force: both kinds of speculation come off seeming, at last, impotent. Rich's protagonist, Mitchell Zukor, reverts to a primal version of himself (once again, learning to die into species being), bulking out and bearding up as he develops basic survival skills, like foraging.

The masculinist Paleo romance that too often serves as the last act of climate fiction is reminiscent both of the survivalist environmental populism of the nuclear age—of which James Dickey's *Deliverance* (1970) offers a respectful critique—and of the popular American Western, where wild (white) men such as John Wayne's Ethan Edwards in *The Searchers* (1956) "tame the wilderness" to make it safe for settlers, which in the Western typically means the nuclear family.

Even a novel as interested in feminism as Theroux's *Far North* offers consistent homage to Western-type scenes, as in the protagonist's concluding advice to her daughter: "When you're ready to light out of here, take the Winchester and the fastest pair of horses and go" (314). Women, at best a silent backdrop to the masculine exertion of the Western, often appear as a site of extreme vulnerability and even as luxury in cli-fi. The female protagonists in Butler's *Parable* series and in *Far North* must pass as men, *The Road* depicts women as slaves and as silent or dead partners, the Finnish author Antti Tuomainen's *The Healer* (2010) is dramatized around the loss of a woman and the near impossibility of heterosexual love in the time of climate collapse, and Edan Lepucki's *California* (2014) explores how the female capacity for pregnancy threatens a climate change survivalist community led by men schooled in Western humanism at an all-male liberal arts college. Even the savvy female survivor of Paolo Bacigalupi's multilayered cli-fi novel *The Windup Girl* (2009) is, after all, a "new person," meaning a kind of cyborg with female-shaped secondary sex characteristics designed for others' pleasure.

The woman problem in cli-fi moves across an ideological spectrum from the popular Western's assumption that new and unsettled environments (in this case, post-Holocene climates) are inhospitable to "vulnerable" female bodies to a more complex concern with population control that is played out through the suppression and even reembodiment of female characters. Haraway's imperative to "make kin not babies!" neatly glosses the ways in which new kinds of affiliation or kin making might assist population control, an issue at the core of her nuanced call to stay with the trouble we humans have brought into being (*Staying*, 101). The highly politicized questions of overpopulation and how to curb it without reprising a colonialist practice perhaps call for a drafting stage through speculative fictions. Margaret Atwood's MaddAddam trilogy (2003–13), speculative fiction with collateral interests in climate change, portrays a sustainable, genetically engineered society in which the cultural significance of gender and sex has dissolved. After a massive die-off of *Homo sapiens sapiens*, Atwood's new humanlike creatures mate as most primates do: infrequently, in the season of estrus, and without anxiety, romance, or art.

**Coda: The Social Work of Cli-Fi, Beginning in Love**

All of this leads to the primary questions haunting cli-fi and other Anthropocene genres invested in imagining the future as we live through the shifting present:

Will there be love in the era of climate collapse? If so, what will it look like? I argue that such questions of intimate attachment, appropriate to the project of the novel since its inception in the eighteenth century, are the ones most likely to drive the quest for a "new humanism," which Scranton and others have endorsed, because love demands, first, a theory of attachment (a fundamental psychological model); second, a practice of caring (a fundamental social practice); and, third, a fundamentally political categorization of worthy objects, of persons—human and nonhuman—deserving of empathy and care. Cli-fi authors have displayed diverse, tentative approaches to the question of love and how it might be answered within nascent climate regimes.

Barbara Kingsolver's *Flight Behavior* (2012), one of the most socially progressive and even hopeful cli-fi novels to come out of the United States, imagines an almost spiritual love (agape) reignited by wonder at other life forms, such as the monarch butterflies who are central actors in her plot. Kingsolver's interest in biophilia as a human expression of attachment that is born of both science and faith suggests the biologist E. O. Wilson's simultaneously empiricist and faith-based arguments for the wonder of biodiversity as an impetus to planetary-scale conservation. In a manner more ironic and polemical than Kingsolver, Jonathan Franzen also suggests in his climate-change-themed essays "My Bird Problem" (2005) and "Carbon Capture" (2015) that love, in the era of climate change, inheres in attachment to other life—but necessarily small attachments to limited life forms and places, attachments that can be enacted through local conservation or appreciation (e.g., birdwatching).[31] The writer Jeff VanderMeer's Southern Reach trilogy (2014), which VanderMeer explicitly asks not be identified with cli-fi but rather with the more capacious category of Anthropocene fiction,[32] offers a posthuman vision of the proliferation of life forms in the wake of human extinction and post-Holocene human evolution. The fertility of the world VanderMeer imagines into being on the U.S. Gulf Coast is both gorgeous and profoundly destabilizing to human bodies. Reading VanderMeer requires working through ecological terror to a non-human-centered and perhaps impossible love for a world that is no longer suited to human thriving.

In a different vein, Dale Jamieson and Bonnie Nadzam's *Love in the Anthropocene* (2015) renders "love" as, among other things, an enactment of memories of the kinds of relationship that the Holocene's climate afforded, such as family trout fishing in mountain rivers. Scranton, too, calls for memory as, if not a mode of love per se, then the most fundamental means of cultural survival, by way of cultivating intergenerational community and knowledge during climate shift. Scranton writes: "If being human is to mean anything at all in the Anthropocene, if we are going to refuse to let ourselves sink into the futility of life without memory, then we must not lose our few thousand years of hard-won knowledge, accumulated at great cost and against great odds" (*Learning*, 109). For Scranton, the humanities are fundamentally a project of shoring up cultural memory and rendering it usable for what cannot be predicted but, in some shape, may have happened before. This practice of respecting the stories, beliefs, technologies, and acts of ancestors resonates with Native Americans' protection of traditional ecological knowledges (TEKs), which have become especially important in the climate change era when traditional food sources and animal migrations are shifting in response to accelerated seasons. As Kyle Whyte reminds us, these TEKs make possible future planning and governmental action for Indigenous peoples.[33]

Embracing some version of indigeneity—living long in a place—has been an appropriative gesture of settler colonialists. Perhaps a better way forward, at a moment when it seems to me that we all must consider carefully what it means to live long in a place, would be to rethink the meaning of "settlement," to move beyond colonialist imagining toward a practice of caring about where it is that we are now. Caring for country, conservationism, loving local and small (as opposed to the view from the airplane) cannot be individualistic practices, and in fact they imply a more ecological and embedded politics. The depiction of these kinds of settled practice, these revised habits, competes in cli-fi with charismatic end-times. Learning to die is a practice of living, after all. And, I reiterate, the novelistic mode always has offered opportunities for trying out and testing material and social relations. The feeling-states that language can evoke through mimicked perception are externalized and simplified in visual narrative.

A gorgeous homage to what I call "deep homing" appears in Richard McGuire's graphic novel *Here* (2014), which recounts the longue durée of a single room from eons prior to its construction (50,000 B.C.E.) to distant futures after what appears to be a climate-induced flood and then curation of the watery waste in a posthuman museum. Perhaps love, in the time of climate change, demands memory *and* speculation, by which I mean attachment to multiple generations, distant futures as well as distant pasts, all times worthy of curation and song. What could be more difficult than loving across time, across futures and pasts not known? Not attempting to do so might be easier, but lonely—especially in the years of early climate shift, our diminished present.

# Notes

1. It is not possible to list every instance of such news, but one example that includes a discussion of the phenomenon of sunshine flooding is Justin Gillis, "Global Warming's Mark: Coastal Inundation," *New York Times*, September 4, 2016, 1.

2. These ideas are developed in many of Stacy Alaimo's articles and books, but for the most succinct version, see Alaimo, "Sustainable This, Sustainable That: New Materialisms, Posthumanism, and Unknown Futures," *PMLA* 127 (May 2012): 558–64.

3. "Transcorporeality" is the key word elaborated throughout Alaimo, *Bodily Natures: Science, Environment, and the Material Self* (Bloomington: Indiana University Press, 2010).

4. I include a long reading of the *New York Times* weather section of November 18, 2012, in my chapter "Humanities after the Anthropocene," in *The Routledge Companion to Environmental Humanities*, ed. Ursula K. Heise, Jon Christensen, and Michelle Niemann (New York: Routledge, 2017), 473–81.

5. Lauren Berlant, *Cruel Optimism* (Durham: Duke University Press, 2011), 6.

6. Kari Norgaard, Living in Denial: Climate Change, Emotions, and Everyday Life (Cambridge: MIT Press, 2011).

7. Alastair Fowler, "The Life and Death of Literary Forms," *New Literary History* 2 (Winter 1971): 199–216, 214.

8. Veronica Hollinger, "Genre vs. Mode," in *The Oxford Handbook of Science Fiction*, ed. Rob Latham (New York: Oxford University Press, 2014), 140.

9. For a more explicit typology see Adam Trexler, *Anthropocene Fictions: The Novel in a Time of Climate Change* (Charlottesville: U. Virginia, 2015).

10. Nancy Armstrong, *Desire and Domestic Fiction: A Political History of the Novel* (New York: Oxford University Press, 1987), 5.

11. Franco Moretti, *Graphs, Maps, Trees: Abstract Models for Literary History* (London: Verso, 2005).

12. Octavia E. Butler, *Parable of the Talents* (New York: Grand Central Publishing, 1998), 8 (hereafter cited parenthetically in text).

13. See Shelley Streeby, "Speculative Archives: Histories of the Future of Education," *Pacific Coast Philology* 49.1 (2014): 25–40; Donna J. Haraway, *Staying with the Trouble: Making Kin in the Chthulucene* (Durham: Duke University Press, 2016), 119–20; Sylvia Mayer, "Genre and Environmentalism: Octavia Butler's *Parable of the Sower*, Speculative Fiction, and African American Slave Narrative," in *Restoring the Connection to the Natural World: Essays on the African American Environmental Imagination*, ed. Sylvia Mayer (New York: Transaction, 2003). Hereafter, all of these sources are cited parenthetically in the text.

14. Olamina elaborates what it means to "shape God" in the first novel: Octavia E. Butler, *Parable of the Sower* (New York: Grand Central, 1993), 24–25.

15. Jameson writes: "The Utopian form itself is the answer to the universal ideological conviction that no alternative is possible, that there is no alternative to the system. But it asserts this by forcing us to think the break itself, and not by offering a more traditional picture of what things would be like after the break." *Archaeologies of the Future* (London: Verso, 2005), 232.

16. Teresa Brennan, *Globalization and Its Terrors: Daily Life in the West* (New York: Routledge, 2003), 19. As the title of Brennan's book indicates, she focuses primarily on the wearing effects of economic globalization in the wealthier regions of the world.

17. Rob Nixon, *Slow Violence and the Environmentalism of the Poor* (Cambridge: Harvard University Press, 2011).

18. Ta-Nehisi Coates, *Between the World and Me* (New York: Spiegel and Grau, 2015), 79 (hereafter cited parenthetically in text).

19. Jerry W. Ward Jr., *The Katrina Papers: A Journal of Trauma and Recovery* (New Orleans: University of New Orleans Press, 2008), 183.

20. Roy Scranton, *Learning to Die in the Anthropocene: Reflections on the End of a*

*Civilization* (San Francisco: City Lights, 2015), 19 (hereafter cited parenthetically in text).

21. Thomas King, *The Back of the Turtle* (Toronto: HarperCollins, 2014).

22. Kyle Whyte, "Our Ancestors' Dystopia Now: Indigenous Conservation and the Anthropocene," in *The Routledge Companion to Environmental Humanities*, ed. Ursula K. Heise, Jon Christensen, and Michelle Niemann (New York: Routledge, 2017), 206–15.

23. Donna Haraway, "Speculative Fabulations for Technoculture's Generations: Taking Care of Unexpected Country," *Australian Humanities Review* 50 (2011), http://www.australianhumanitiesreview.org/archive/Issue-May-2011/haraway.html.

24. The ambivalent anticolonialist interplanetary settlement fantasy of the *Parable* novels does not quite reconcile with my notion of the everyday Anthropocene. It is more usefully considered within the cultural moment of Afrofuturism in the 1990s. Alondra Nelson, a primary generator of Afrofuturist thought, notes that "future vision is a necessary complement to realism, for the reality of oppression without utopianism will surely lead to nihilism." Nelson, "AfroFuturism: Past-Future Visions," *Colorlines* 3 (Spring 2000): 34.

25. J. G. Ballard, *The Drowned World* (1962; repr., New York: Liveright, 2012), 62.

26. Dipesh Chakrabarty, "The Climate of History: Four Theses," *Critical Inquiry* 35.2 (2009): 197–222; Chakrabarty, "Brute Force," *Eurozine*, October 7, 2010, http://www.eurozine.com/brute-force.

27. Cormac McCarthy, *The Road* (New York: Vintage, 2006).

28. George Monbiot, "The Road Well Traveled," *Guardian*, October 30, 2010, http://www.monbiot.com/2007/10/30/the-road-well-travelled.

29. Emily St. John Mandel, *Station Eleven* (New York: Vintage, 2014), 32 (hereafter cited parenthetically in text).

30. Marcel Theroux, *Far North* (New York: Picador, 2009), 261 (hereafter cited parenthetically in text).

31. Jonathan Franzen, "My Bird Problem," *New Yorker*, August 8, 2005; Franzen, "Carbon Capture," *New Yorker*, April 6, 2015.

32. Jeff VanderMeer's thoughtful reflections on the genre terminology appear in his blog of March 16, 2016: "Global Warming Narratives: The Dangers of Pushing for Early Labeling," http://www.jeffvandermeer.com/2016/03/16/global-warming-narratives-the-dangers-of-early-labeling.

33. Kyle Powys Whyte, "What Do Indigenous Knowledges Do for Indigenous Peoples?" in *Keepers of the Green World: Traditional Ecological Knowledge and Sustainability*, ed. Melissa K. Nelson and Dan Shilling (forthcoming), manuscript p. 14.

# 13.
# Ungiving Time
## Reading Lyric by the Light of the Anthropocene

*Anne-Lise François*

Sweat turned into a cloud that fell as snow to finally become ice
—Camille Seaman, www.vanishing-ice.org

In Northamptonshire and East Anglia "to thaw" is *to ungive*. The beauty of this variant surely has to do with the paradox of thaw figured as restraint or retention, and the wintry notion that cold, frost and snow might themselves be a form of gift—an addition to the landscape that will in time be subtracted by warmth.
—Robert Macfarlane, *Landmarks*

In her paper "Feral Biologies," Anna Tsing claims that "the inflection point between the Holocene and the Anthropocene might be the wiping out of most of the refugia from which diverse species assemblages (with or without people) can be reconstituted after major events (like desertification, or clear cutting, or, or, . . .)." For Tsing, "the Holocene was the long period when refugia, places of refuge, still existed, even abounded, to sustain reworlding in rich cultural and biological diversity."[1] In other words, what permanent climate instability threatens is the existence of margins as an alternative way of accomplishing x.

The margin can be understood both temporally and spatially as the capacity to alternate, as when temporary riverbeds shift between dry and wet land or fields alternate between lying fallow and being cultivated. Tsing may also be understood as designating the Anthropocene as a time of intensified marginality, if we remember that in ecology, the margin or boundary zone carries a related but different sense of the limit of a given plant species' viability—the edge where it is still possible, but only just, for certain varieties, but not all, to survive and where normally insignificant differences in genetic plant races and environmental conditions begin to matter, because plants no longer have the ability to make up for the deficiency of x (or to confront y) by availing themselves of z. Here is the agronomist Wes Jackson citing biologist Richard Levins's summary of what is known as Schmalhausen's law, after the Russian ecologist Ivan Ivanovich Schmalhausen: "Organisms near the boundary of their environmental tolerance or geographic range are very sensitive to local conditions. Trees that grow on any soil type in most of Eurasia are limited to limestone soils near their northern boundary. Soil differences that did not matter in Central Russia were a matter of survival at the edge of the tundra."[2] The margin in this sense can be thought of as the spatial equivalent of the terminal futurity inhabited by what ecologists call "ghost species"—species that, having established themselves under certain environmental conditions, persist without the ability to spread or reproduce themselves once those conditions have changed.

The concept of the margin, then, works antithetically to name both an extremity and the measure that might cushion it. Compare, for example, Levins as quoted by Jackson—"at the extreme edges of tolerance, almost anything can destabilize a system.... It is the marginality itself which makes people vulnerable to trivial differences of experience, skill, energy, acuity or inclination, almost making perfection a necessity and loading choices with grave consequences" (*Altars*, 144)—to the salutary sense that Raymond Williams gives the margin as a buffer, however thin, between total dependence on wage labor and a measure of autonomy when recalling his father's allotment garden:

> Again and again, down to our own day, men living in villages have tried to create just this kind of margin: a rented patch or strip, an extended garden, a few hives or fruit trees. When I was a child my father had not only the garden that went with his cottage, but a strip for potatoes on a farm where he helped in the harvest, and two gardens which he rented from the railway company from which he drew his wages. Such marginal possibilities are important not only for their produce, but for their direct and immediate satisfactions and for the felt reality of an area of control of one's own immediate labour.[3]

In either case, whether we understand the margin as a principle of deliberate inefficiency, redundancy, and noncommitment to an exclusive way of doing things, or as a name for a ghostly outliving without the means of retreating from the edge back into the general pool, the Anthropocene might be no more than a name for whatever within capitalist modernity forces the definitive foreclosure of other ways of being (or other "uses" of the same space by other creatures).

In December 2015 at the workshop "Climate Change and Its Challenges to the Scholarly Habitus," co-organized by anthropologists from the Institut de Recherche pour le Développement and from Johns Hopkins University on the occasion of the COP21 talks, I heard geographer and historian Anupam Mishra put the difference between centuries-old traditional water collection methods in India and twentieth-century underground pipelines in terms of the few drops that you might let fall when taking water from the former, thereby inadvertently satisfying the thirst of nearby birds. Simultaneously, in an installation at the Place du Panthéon timed so that the ice would be quietly melting in the course of the COP21 talks, the artist Olafur Eliasson had arranged in the form of a clock dial twelve blocks of ice from melting icebergs that he had fished out of the ocean's salt waters. In the photo sent to me by Rochelle Tobias is just discernible, in her words, "a drop falling from one of those evanescent but also very present ice sculptures" (pers. comm.).

If all we knew of Eliasson's installation was an image of a lonely drop of potable water disappearing untapped, then, whatever the irony of the artwork's costs of production (including shipping the ice to Paris using the same fossil fuels responsible for rising temperatures), the ice clock might be thought of as an elegiac gesture that does nothing to stop or to save. It only registers the letting go as sweet water melts [into salt]. At most, the work momentarily suspends the will to extraction and conversion into immediately usable, disposable but also storable energy, a will whose very success has brought us to the nightmare point now trackable in the accelerated melting of the world's glaciers.

So at the close of "After great pain a formal feeling comes," Emily Dickinson registers the moment of giving up when the hands open again and the clutching ends:

This is the Hour of Lead -
Remembered, if outlived,
As Freezing persons, recollect I the Snow -
First - Chill - then Stupor - then I the letting go -[4]

The irony of this final, riddling simile defies parsing. The hour can only be remembered if outlived, but the energy it takes to survive might leave one bereft of the power to engage in the reflective activity of remembering. If the hour of "formal feeling" that followed "great pain" (the aftermath of consciousness) is remembered—because outlived, because put behind one as something definitively past—it is only by analogy to how "Freezing persons, recollect the Snow," the stage before freezing when death has not hardened into the inexorable. If we assume that the "Freezing persons" are those who freeze to death, then the image by which Dickinson figures the prospective act of retrospection—the glance that lucky survivors might one day cast back on a final hour when it has been "outlived"—is that of a dying person's last thought. Or the freezing persons might be imagined as markers in an otherwise undifferentiated landscape. Since snow in Dickinson's other poems refers to the decreative artistry that erases human divisions, to "recollect" the snow might be to gather again that which has been dispersed and released, or allowed to fall indiscriminately.[5]

As Sharon Cameron notes in her still unsurpassed reading of the poem, the progressive hardening from "wood" to "stone" to "lead" that, in the earlier stanzas, tracks the disappearance of a subject who would be there to register the pain of loss, reverses course in the final stanza, as the strange fixity of the aftermath—a time (it would seem) to which nothing more could happen—gives way to "a more complex [image]," marked by "its susceptibility to transformation, its capacity to exist as ice, snow, and finally as the melting that reduces these crystals to water."[6] For Cameron in *Lyric Time*, the last line's "undoing of the spell of stasis" (168) marks a final defeat of the Dickinsonian (and lyric) project of forestalling transience and hanging on to loss as the trace of, and impossible substitute for, the lost object: "In the sequence of diminishing returns, what has been is, by definition, missing. What remains is a true blank, the genuine space at the thought of which despair 'raves—,' and around which words gather in the mourning that is language" (169). For Annie Finch, too, the last two lines' resolution into iambic pentameter marks a surrender, an abdication of the declarative autonomy that Dickinson is able to achieve in the shorter common meter lines (*Ghost of Meter*, 29–30). Yet the last line also overlays a syntactic patterning across the iambic pentameter's alternation of weak and strong beats:

First - Chill | / then Stu | por - / then | the let | ting go -
w         S       w       S       w        S       w  S   w  S

Here the vertical bars separate the five metrical feet, while the slashes and different font sizes indicate the lengthening syntactic units. Each stage, like the fingers of an unfolding hand, or like the confidence that grows in the act of finding one's way, is one, then two, syllables longer than the last, in an accretion that signifies the release by which the hour of lead expands to include its own dissolution. Something familiar enough to call "nature" or natural time, some sense of a reliably predictable temporal order by which to orient oneself and navigate a course reenters the poem here in the relatively ordinary sequencing "first x, then y, then z"—even if too late for those lost in the undifferentiated winter landscape. Within the poem this hint of a release back into successive time would be the only grounds for the utopian reading I wish to give of Eliasson's melting ice, as of a redemptive release or letting go of the will to hang on hard to things.

I offer this brief opening excursus as a conceptual map of what I attempt in this chapter, which is to read poetry and discourse on the Anthropocene in relation to one another, as attempts to hold together overlapping yet semi-autonomous temporal scales. As a figure for the collapse of slow time into fast, how can the Anthropocene help us consider anew the problem of lyric time as distinct from empirical time? How might attention to the collapse of millennia-long temporal accumulation into decades-short meteoric consumption help us remediate certain common motifs of the lyric as the "intensification of presence" (Miner) or as the "flash of the momentary against the enduring" (Backus) or as a "moment's monument" (Rossetti)?[7] To what extent does the formalization of scale variance and of the play of different techniques for condensing and extending, slowing and accelerating time, which goes by the name of "lyric," anticipate and even neutralize the shock of the Anthropocene? The ontic "fact" of time's passing may never have been as self-evident as supposed, but nothing is now less certain than its linear diminishment and loss without return; nor can one now simply assume the "direct reversion of insensate world to natural, uncomplicated plenty" (Cameron, *Lyric Time*, 216). Gone is the automatism of the fallen "nature" that Cameron sees Dickinson's poems as struggling to deny and overcome.

The melting of Eliasson's ice blocks is a figure for the collapse of crysopheric deep time into human experiential time, which for Dipesh Chakrabarty spells the end of the division between the natural and human sciences. In his remarks at the Paris workshop, the anthropologist Philippe Descola reminded us of the need to think the difference between the continuous "anthropizing" (or mutual acculturation) of the planet—stretching over fifty thousand years—in more or less regular rhythms of alternation between clearing and reforesting, gardening and rewilding, and the Anthropocene, usually conceptualized as an abrupt and irrevocable shift. The paradox waiting to be thought in the idea of the human species suddenly acquiring the power to affect the Earth system consists in precisely this suddenness, in the simultaneity of speed and slow time, or the disjuncture between the enormity of the consequences—as if for the rest of time to come—and the relatively brief parenthesis, a fraction of a second in evolutionary time and a blink even on the scale of human history. To listen to the geologists, the Anthropocene would be humanity's last cigarette, a name for the fast consumption of deep time—a short hour of *consommation* in the double sense of consumption and consummation (as if you could burn millennia in the time of a cigarette or smoke deep time), framed on either end by deep time: the millions of years during which concentrated solar energy was captured and accumulated, and the eons still to come, yet already concluded, blocked by the indelible consequences of their release. A "fleeting folly" monumentalized for time to come is how the geologist David Archer puts it in *The Long Thaw: How Humans Are Changing the Next 100,000 Years of Earth's Climate*: "Mankind is becoming a force in climate comparable to the orbital variations that drive glacial cycles. The long lifetime of fossil fuel $CO_2$ creates a sense of fleeting folly about the use of fossil fuels as an energy source. Our fossil fuel deposits, 100 million years old, could be gone in a few centuries, leaving climate impacts that will last for hundreds of millennia. The lifetime of fossil fuel $CO_2$ in the atmosphere is a few centuries, plus 25% that lasts forever."[8] Yet the Anthropocene-as-last-cigarette is a trope that is not content-specific, a point worth stressing given the tendency in mainstream environmentalist discourse to fixate on fossil fuels in a way complicit with the market logic that seizes on carbon emissions as abstract, quantifiable, and tradable. The extraction and conversion of uranium into nuclear energy carries the same dialectic of heavy liquidity.[9] As I note further below, the recurring motif in Anthropocene discourse of the prodigal squandering of geological time echoes the figure of the reckless expenditure of soil fertility found in critiques of conventional agriculture (understood as the planting and replanting, year after year, of annual crops), as if carbon-based capitalist economies were only the magnification of a structure of borrowed time already lightly inscribed within the Neolithic revolution.

Once in a collaborative, cotaught research seminar on the Anthropocene, a graduate student, Joe Albernaz, proposed that the Anthropocene is a catachresis, a figure for something that there is no direct way of naming or referencing.[10] The favored example of a catachresis among Derrideans is Aristotle's metaphor for an action that would otherwise remain "nameless"—"the sun sows its rays"—a beautiful image for the way the Sun distributes its energy, or allows it to be pocketed, more or less evenly, in the course of a single day."[11] The figure of "sowing"—a kind of metalepsis or substitution of anticipated effect for cause—evokes the role of solar energy in plant growth and by inversion the role of plants in converting the Sun's rays into usable energy through photosynthesis. Amid cascading reports of monthly heat records, there occurs another misrecognition of "inside" and "outside," as of effect for cause: the heat we might attribute to the Sun's "rays" is in fact the palpable expression of the greenhouse gases in the Earth's atmosphere.

Aristotle's example of the Sun sowing its rays acquires yet another valence when set alongside Andreas Malm's revision in *Fossil Capital* of historian E. A. Wrigley's claim that the Industrial Revolution represents a break from dependence on the "yield of present photosynthesis."[12] Until then, according to Wrigley, humans had been limited to what the land could supply in real time—trapped, as it were, like Nietzsche's animals within the relative present of the Earth's annual rotation around the Sun. But "when iron, pottery, bricks, glass, salt and other industries turned to coal, they bypassed the restricted surface area by digging into the stores of *past* photosynthesis, wholly new vistas of expansion opening up beneath the forest and the field" (Malm, *Fossil Capital*, 21). Rejecting as inaccurate Wrigley's "organic" versus "nonorganic" distinction, Malm instead divides the energies ultimately derived from solar radiation into categories of "flow" (wind and water); "animate power" (human and animal labor); and "stock" (coal and other fossil fuels). "Requir[ing] no special human power to bring them forth"; "caught for an instant as they passed by"; and "slipp[ing] away" "as soon as they had been harvested" (39), it is easy enough to recognize the Heraclitean dimensions of wind and water as Malm describes them: there for the taking but also nonretainable. As a figure for capital itself, the category of "stock" represents, by contrast, something more vexed: the site of a double retention and release, a release all the more illusory for only initiating another cycle of accumulation and deferral.

Remembering that agriculture too is often credited with instituting a shift from living within the constraints of the moment to tapping into deep time, we might pause before deciding on what changes in humanity's temporal ontology with the sudden availability of the accumulated energy of thousands upon thousands of solar years. Indeed Wrigley's historical argument about a shift from dependence on (and movement with) organic energy cycles to a more perpendicular inorganic energy extraction is easily relativized if we remember that well before coal mining, conventional agriculture borrows from the deep past to extend the storage life of otherwise perishable energy supplies. In Wes Jackson's words, "traditional agriculture coasted on accumulated principle and interest, hard-earned by nature's life forms over those millions of years of adjustment to dryness, fire, and grinding ice" (*Altars*, 80). Yet the annual planting (or replanting) of crops whose seeds are harvested as grain rather than allowed to propagate also entails, as Colin Duncan has argued, a constant struggle against encroaching life forms and a tremendous output of energy to keep the land "free," so that each year the biotic community has to start again at zero rather than attain maturity (or evolve to something else).[13] Ransacking the very old underneath for the sake of maintaining the illusion of perpetual youth at the surface, as if in a kind of earthly rewriting of *Dorian Gray*, annual agriculture, on this account, cuts against and into the course of time at both ends in a disturbance that freezes, as much as it accelerates, the Earth's relatively more natural flows of energy.

But what would be the alternative to "coast[ing] on the sunlight trapped by floras long extinct," as "modern agriculture," according to Jackson, does (*Altars*, 80)?[14] The normative ideal of living within one's means (as elusive as the dream of living in the moment and only on what the moment avails) suggested by Jackson's implicit metaphors of credit and debt would seem to leave little room for acknowledging non-self-sufficiency and for affirming intergenerational and interspecies dependence. A common move in the bioregionalist strain of first-wave American environmentalism was to define living in relation as living within the limits of a given watershed—testing those limits, stretching and extending them, but never too far. One could similarly expand the temporal framework of the "present" to allow for the arrival of long-delayed *lettres en souffrance* (letters in transit) and to count among the riches of the "now" their deferred return, in accordance with the principle of deferral as circulation, seen in many gift economies.

Assuming it were even possible to compare rhythms across such large swaths of human history, at stake in the kinds of divisions that Wrigley, Malm, Jackson, and Duncan are making is the nature of the human accompaniment—more or less in synch? more or less contrapuntal? or simply and stupidly indifferent?—to otherwise autonomous cycles, orbits, and still unfolding temporal processes of accretion and diminishment. As Dickinson writes, "the stiff Heart questions 'was it He, that bore,' - / And 'Yesterday, or Centuries before?'"

In *Capitalism in the Web of Life*, Jason Moore dismisses Malm's work for "reproduc[ing Malthus's] original error" of "taking the dynamics of nature out of history" and leaving unchallenged the familiar story of a subject called "humanity" whose needs and desires are posited as limitless, except insofar as bounded by the nonnegotiable limits of "nature."[15] Because these limits are erroneously taken as "external" rather than coproduced, nature's autonomy in this narrative is only ever encountered as an unhappy contingency or constraining limit, if not as an openly antagonistic, niggardly, withholding force. But Malm's richly detailed historical account has the advantage of identifying the specific material properties of coal that determined, without rendering necessary or inevitable, its adoption over wind and water in early nineteenth-century industrial Britain. This account helps articulate the temporal logic that, by extracting from and rousing the dormant past so as to indefinitely extend the present, produced the peculiar combination of dormancy and availability that Heidegger would later call *Gestell*. The specific conditions reviewed by Malm include the necessity of extracting coal, before its conversion into energy, from the deep tombs below ground in which it would otherwise have lain lost; the presence of a ready pool of human labor available to do this work; and coal's consequent ability, once disembedded, to circulate on the market as a commodity, features at the time inapplicable to water and wind. Here, in emphasizing how the illusion of coal's freedom and mobility depends on a different kind of freedom—that of the workers whom Marx called "free" when they had been mobilized, disinterred from their contexts, and robbed of independent means of subsistence—Malm rejoins Moore, who also argues that capitalism depends as much on the appropriation of the unpaid work of human and nonhuman agents "outside the commodity system" as on the exploitation of wage labor. Comparing the use of the "outside" in the following two passages, we might conclude that if fossil capital constitutes a determinate shift in humanity's relation to deep time, it would lie in the conversion of stock (in the sense of accumulated funds) to disposable cash, ready reserves:

> In dormancy, *outside* the landscape, the stock could be transformed into an actual source of energy only through massive inputs of human labour, mobilised in a long chain stretching from deep inside the mine. (Malm, *Fossil Capital*, 91, emphasis added)

> Appropriation, in what follows, names those extra-economic processes that identify, secure, and channel unpaid work *outside* the commodity system *into* the circuit of capital. (Moore, *Capitalism*, 17, emphasis added)

In both cases, the awkwardness of the spatial metaphor indicates the double inclusion and continued exclusion of the deep geological past—the still-dormant temporal "outside" within—that when abstracted as pure potential makes possible capitalism's uninterruptible, sleepless, and ceaselessly available present. Here we can see just how continuous schemes for carbon emissions trading and carbon sequestration through forest plantations remain with the logic of treating coal as a capital, an atom of potential energy or unit of potential value, in contrast to the wave or current that is lost as it passes.

Returning now to the notion of the lyric as a technique of presencing, intensification, and condensation—indeed, sequestration—of linguistic utterance into a single verbal image on the page, we might reconsider David Archer's claim, "Mankind is becoming a force in climate comparable to the orbital variations that drive glacial cycles" (*Long Thaw*, 11), in relation to Dickinson's "Safe in their Alabaster Chambers." In the poem's first stanza Dickinson describes the dead in their coffins as those who have turned their faces away from the temporal, have given up all things terrestrial, and now await release from mortal time into eternity. Or rather, having given up even waiting, they are suspended between the mortal time of succession (in which one thing may follow another) and the release of eternity, which is why their exposure to the divine promise of redemption looks like enclosure in the smallest temporal space imaginable:

> Safe in their Alabaster | Chambers -
> Untouched by Morning -
> And untouched by Noon -
> Lie the meek members of | the Resurrection -
> Rafter of satin - and Roof of stone!
>
> (1.162)

Remembering the silenced chant "Keep It in the Ground" with which Paris's emptied streets should have been resonating during the COP21 talks, it is difficult not to read Dickinson's image of burial or safekeeping through the lens of the Anthropocene. Whatever their reference to the Christian promise of a resurrection to come, her lines ask us to imagine an already achieved and immanent intactness, and isn't this also what the activists are trying to envisage—a burial not for the sake of a moment of use to come nor to get rid of something unwanted (as in the case of radioactive waste), but as a kind of immanence, already achieving what it desires, not a deferred wealth waiting to be retrieved?

The challenge here is to articulate the conceptual—as well as the more obvious political—difference between carbon sequestration, as promoted by finance capitalism, and the active nondevelopment of what already lies underfoot, as called for by many Indigenous peoples and climate justice activists. To the extent that "climate-conscious" forest management often favors uniformly aged, grown, and logged tree plantations, carbon sequestration represents a deliberate stoppage of ecological time while nondevelopment expresses a deliberate choice to pass on and let continue processes that, however autochthonous, only minimally coincide with human temporalities.

Teasing us with the ambiguity of safekeeping exhausted remains that may yet potentiate another time, "Safe in their Alabaster Chambers" offers an alternative to the false choice between radical noncoincidence and the pseudo inclusiveness of integrating forests into the global market as carbon stocks that are in one sense permanently saved—they can never stop breathing or doing the breathing for us—and in another sense precariously temporary, able to be replaced at any time. In Dickinson's poem, a dependence on rescue from beyond maintains a passivity so still it looks like self-sufficiency: if the "meek" (whom it is impossible simply to designate as dead) are those who can do nothing more *for* themselves, the initial downbeat falls on "Safe" and on the sense that nothing more can be done *to* them.

In her French edition of the poems, Françoise Delphy departs from previous translators in translating "safe" as "à l'abri" instead of "en sûreté," a choice that has the ironic effect of shifting the poem to a pastoral key.[16] Where "safe" seals in stone, "à l'abri" suggests a sheepfold, a temporary and more or less permeable shelter from wind and rain. Etymologically, "abri" derives from the Latin *apricor*—to sun oneself—so that a retranslation would render the extraordinary image, "In the sun . . . untouched by morning / and untouched by noon." Depending on how we tally the Sun—as the purveyor of a day's worth of energy, of a year's, or of millennia's in the form of fossil fuels—pathos here lies in the changing tone of a phrase that once promised protection and now spells disaster. But Dickinson is also tracking its mounting course metrically and, whether we carry over the first line's trochaic pattern and let the downbeat fall on the "un" of "untouched," or hear the second line as an iambic interval, something shifts in the very recording of nonchange, between the first and second iteration of "untouched." The hands of the poem's metrical clock advance as if in the dark of the tomb by a kind of vocalized touching—the patterning of off and on beats—to record a first and then a second noncontact. Just discernible in the recognition of something unshareable within otherwise momentarily shared weathers is an alternative to the easy convertibility of x into y and back again assumed by the logic of carbon emissions trading.

But it is the stanza that follows that appears to anticipate Archer's attempt to capture through metaphor the strangeness of coordinating a "small" subject, such as "mankind," with the orbital variations of glacial cycles:

> Grand go the Years - in the | Crescent - above them -
> Worlds scoop their Arcs -
> And Firmaments - row -
> Diadems - drop - and Doges - | surrender -
> Soundless as dots - on a | Disc of snow -
>
> (1.162)

As I argue in *Open Secrets*,[17] the first four lines make a dactylic waltz of war in heaven, collapsing with arrogant nonchalance historical and celestial or planetary time: from the impossible perspective of the dead below, galaxies spin as fast as heads roll and cities fall. Polysyllabic splendor, to borrow David Porter's term,[18] combines with synecdochal compression—the cut of "diadems - drop" elides even the heads that wore them—to lend erotic charge to the "row" of "Firmaments," which it is hard not to hear according to the other sense of "row" from Dickinson's "Wild Nights" ("Rowing in Eden - / Ah - the Sea! / Might I but moor - - tonight - / In thee!"; 1.288). But something else happens in the transition from the first four lines to the fifth: history, whether planetary or global or metropolitan, comes down and falls back into the terrestrial, and the cosmic condenses into something touchable or into something you can imagine holding in one hand, however tenuous—indeed, untenable and incapable of bearing weight—the image of soundless dots on a disc of snow. As Porter suggests, the image may be inscrutably abstract in the sense of difficult to visualize or read in narrative terms, but the line, which Dickinson sometimes copied as one and sometimes lengthened to two, as if enacting the drop it describes, nevertheless makes something available to sensory perception in its chiastic patterning of the "s"s and "d"s of the previous line's "surrender" (Porter, *Dickinson*, 26). The line flowers in the expansion of monosyllables, or something opens in this final meting out of stops and sibilants, yielding a kind of surrender of surrender itself, beyond erotic release, and giving us back, in the image of snow falling on fallen snow, one dimension falling on two, or the thinnest of particles holding and then yielding under its own weight, the experience of time passing or temporal disappearance. In this sense, however much, as Porter rightly suggests, the snow "inexorably deprives 'safe' of its coziness" (26), this reexposure or unsealing also puts us back in time in a way that recalls the downward, earthbound movement of epic as it turns to pastoral, according to one common narrative of Miltonic and post-Miltonic Romantic poetry. Or to trace the line of literary descent back even further, Dickinson, who declared "I think that the Root of the Wind is Water" (3.1123), helps make newly legible the sixteenth-century song "Western wind" and in particular why the metrical weight should fall on "down" in the line "the small rain down can rain." For fundamental is the wish not just that it would rain but that there be a place to receive it, that it might fall even on itself.

To return, then, to the question of what it means to read lyric poetry in the Anthropocene: Is there more to say than that, in the suddenness with which they can shift gears without it feeling like a shock, poems can help make palpable the contradictions of the simultaneously fast and slow times of the Anthropocene? Two figures common to lyric—parataxis (the minimalist practice of placing side by side two syntactic phrases without subordinating one to the other or explaining the connection between them) and succession (the temporal doubling of near identical verbal patterning by which poems often map human passing onto seasonal passage)—emerge in particular here as two instruments for registering both scalar transference and nontransferability and for indicating both the simultaneity and noncoincidence of different tempos of change and mortality. Parataxis often proceeds by recession, paring down the two terms or phrases it conjoins while also retreating from the work of persuasion, so that whatever is made available is also an occasion not to take it up. Journalistic and academic accounts of climate change also often present the rapid recession of glaciers as an occasion for revelation by subtraction—for the distillation of truths and the dissolution of temporal illusions, illusions in particular about the ability to distinguish near and distant future and past. So a *Smithsonian* article announcing that a buried Cold War, radioactive U.S. military base will soon emerge from melting ice in Greenland reads as an exercise in telescoped time, the ghastly inversion of an investment that was not supposed to mature for centuries (or a debt to amortize) suddenly returning on the hour or flowering at one's feet.[19] Similarly, accounts of how climate change will affect pollinators and the flowering species that depend on them often focalize the disaggregation of temperature and light—temporal cues that once coincided and now no longer do—in ways that repeat what can happen when a poem presents, and then dispels, the semblance of mutual accord as merely passing conjuncture.

I would like to close by juxtaposing two poems lifted irresponsibly from the contexts in which they are usually read, with the thought that this disembedding is partially justified in a climate defined by competing allegories of disinterment and sequestration. Both texts hang on the analogy between the poem that changes hands and human hair that, all but weightless, can be cut and circulated. If in both cases this gift is turned down or only ambiguously taken up by indirection, there is also the suggestion that this capacity to set something down, or to be set down, might be the (vanishing) gift in question.

Paul Celan's early poem "Die Hand voller Stunden" (Your hand full of hours) consists of a series of non-exchanges between a woman who makes these offerings (and advances) and the male speaker who declines them. In the "hollow way" between the two lies the unspoken knowledge whose weight the poem bears—knowledge of the recent events of human history, including the grotesque stockpiling and rendering of the hair (presumed brown but not always brown) of the victims of the Nazi death camps, a rendering expressive of a will to efficient extermination so mad that it would wipe out even the last trace of something supplemental or left over, and knowledge of literary history, of the figural role played by Laura's hair in the Petrarchan poetic tradition, as that whose *oro* is there to be lifted by the *aura* (or winnowing wind). The impossibility either of keeping these two histories separate or of making them commensurate with each other accounts for the pathos in the lightness with which the addressed you is supposed to have lifted the hair onto the scales of grief (the German is *aufheben* with all its many meanings of retention and release):

> Die Hand voller Stunden, so kamst du zu mir - ich sprach:
> Dein Haar ist nicht braun.
> So hobst du es leicht auf die Waage des Leids, da war es schwerer als ich ...
>
> Sie kommen auf Schiffen zu dir und laden es auf, sie bieten es feil auf den Märkten der Lust -
>
> Du lächelst zu mir aus der Tiefe, ich weine zu dir aus der Schale, die leicht bleibt. Ich weine: Dein Haar ist nicht braun, sie bieten das Wasser der See, und du gibst ihnen Locken ...

Du flüsterst: Sie füllen die Welt schon mit mir, und ich bleib dir ein Hohlweg im Herzen!
Du sagst: Leg das Blattwerk der Jahre zu dir—es ist Zeit, daß du kommst und mich küssest!

Das Blattwerk der Jahre ist braun, dein Haar ist es nicht.

Your hand full of hours, you came to me—and I said:
Your hair is not brown.
So you lifted it lightly on to the scales of grief; it weighed more than I . . .

On ships they come to you and make it their cargo, then put it on sale in the markets of lust—

You smile at me from the depth, I weep at you from the scale that stays light.
I weep: Your hair is not brown, they offer brine from the sea and you give them curls . . .
You whisper: They're filling the world with me now, in your heart I'm a hollow way still!
You say: Lay the leafage of years beside you—it's time you came closer and kissed me!

The leafage of years is brown, your hair is not brown.[20]

Michael Hamburger translates "das Wasser der See" as "brine from the sea," no doubt for the sake of the internal rhyme with "brown" and perhaps for the association of brine with bitterness, tears, and even sex. But this translation produces a humanist reading registering shock at the prostitution of the woman's curls for the worthless abundance of what the venal offer up in exchange, as if she were the bearer of unique, nonsubstitutable value, whereas, on the contrary, her hair is defined in relation to what it is not and for which it might but does not substitute—brown. In any case, an Anthropocene reading of the poem would no longer be able to take for granted either the cheapness of salt water or the depthlessness of the merchant seas. The collapse of deep geological time into the present-day carbon and uranium extractive economies would also give a different valence to the poem's close, where the woman invites the man to meet (answer or double) her act of putting the hair on the scales of grief by laying beside him the leaf work of years, whether to set it aside for her

Ungiving Time

or adorn himself with it. Inverting the first line's "kamst du zu mir," "Leg ... zu dir" leaves ambiguous the verb's relation to the dative "dir" as can be seen from the line's various translations: "lay ... beside" (Hamburger); "mets avec toi" (put with you) (Lefebvre); "lay down" (Weinfield). Is her request that he assume or put down the leaf work of years gone by? And is this burden comparable to the hair in its heaviness or in its seeming lightness as ornamentation—as the leaves through which light passes and as the pages in which rustle the words of earlier poets (if we remember the art historical sense of "Blattwerk" as the monumentalization of the organic in stone, as well as the pun that Celan is probably making on "leaf work" as the work of poetry)?[21]

The exchange appears to dramatize the relatively familiar ethical choice between keeping faith with the mutilated and dead or letting go of the past to accord the present its due.[22] But what can "es ist Zeit" mean, and what becomes of such a choice in an age of trapped carbon when the "years" no longer give way, only compound, magnify, intensify their effects? A *Smithsonian* article from 2014 reports that decomposers—the fungi, insects, and microbes that usually complete the woodland cycle of decay—have significantly declined in the forests around Chernobyl such that leaf litter is now accumulating on the forest floor, slowing the growth of trees and contributing to high fire risk. Experimenting with bags of leaves deposited in different areas, researchers found that leaves left in areas of high radiation still retained "60% of their weight" after a year, in contrast with other areas, where 70–90 percent of the leaves were gone in that time.[23] What would it mean to, or who would be able to, lay aside or put on these accumulating, nondecomposing leaves?

With this image of evenly distributed experimental bags of matter awaiting decomposition in the now time-stopped (emptier yet still self-filling) woods, I turn to one final text, from Robert Hass's translation of Matsuo Bashō:

My brother opened a keepsake bag and said to me, "Pay your respects to mother's white hair. Now, your eyebrows look a little white, too."

It would melt
in my hand—
the autumn frost.[24]

As with the exchange between the man and woman in Celan's poem, in place of dialogue all we get here is what Roland Barthes would have called "la réponse à côté" by which Bashō sidesteps his brother's demand to confront head-on their shared mortality and inevitable passing.[25] Reception and refusal are as ambiguously linked here as in Celan's poem: it is unclear whether the *hokku* with which Bashō answers his brother's words meets the demand that he honor their dead mother by placing himself in her lineage, or declines to take what the brother wishes to put in his hands. The lock of white hair is a synecdoche for the bag of keepsakes from which it is taken: a sign of mortality that has also been removed or taken out of mortal time, a memento mori that survives the very forms of transience it is meant to provoke the living into remembering. Bashō answers with an image of another kind of lapse or transience, as if to remind himself and his brother that the lock of white hair is like the autumn frost, still subject to further metamorphosis and disappearance.

In *Bashō and His Interpreters*, Makoto Ueda provides the full text of the *haibun*, which appears in Bashō's *Journal of Bleached Bones in a Field* (1684–1685), along with a transliteration of the Japanese, a word-for-word translation, and a further explanatory note:

> I arrived at my native town at the beginning of the ninth month. Nothing of my late mother remained there anymore. All had changed from what I remembered. My older brother, now with white hair in his side-locks and wrinkles around his eye-brows, could only say, "How lucky we are to meet alive again!" Then he opened a keepsake bag and said to me, "Pay your respects to Mother's white hair. They say the legendary Urashima's hair turned white the instant he opened the souvenir box he had brought back from the Dragon Palace. Now your eyebrows look a little white, too." We wept together for some time.
>
> should I hold it in my hands / it would melt in my tears /—autumn frost
>
> te | ni | toraba | kien | namida | zo | atsuki | aki | no | shimo
> hand | in | if take | will-vanish | tear | ! | hot | autumn | 's | frost
>
> Urashima was the young hero who visited a Dragon Lady's palace under the sea. Returning to his native village and finding nothing there that he could remember, he disobeyed the lady's order and opened a jewel box she had given him. Instantly he turned into an old man.[26]

The comparison shows that Hass omits the reference to the warm tears, leaving the reader to infer that the speaker's hand would by its warmth melt the object it might hold, whether hair or frost. The blood that circulates in the living hand—also the source of the saline tears—is sufficient to indicate the poet's responsiveness to his mother's death and to his and his brother's loss, as if warm-bloodedness were the condition not simply of being alive but of being capable of being moved to tears. By paring down what it receives and retransmits of Bashō's verse, Hass's translation renders even more delicate the work of inference that Bashō asks of his reader; by identifying reduction with reception, the translation enacts at the metalevel the melting described.

Loss, disappearance, erasure, survival, displacement, metamorphosis. Bashō's verse and the stories it weaves together—the encounter with the brother and the parable of Urashima—along with Hass's translation are newly legible in a world increasingly defined not simply by the age-old problem of forgetfulness, but by too much retention, too much accumulation, too much permanence. While the climactic changes now grouped together under the term "Anthropocene" will likely spell the end of conditions favorable to life for most humans and other familiar life forms, they also signal what we might call the Earth's monumentalization—or encryption—the moment at which it assumes the work, once assigned to culture, of archiving human history by acting as the storage house or "mansion" or "jewel box" for non-erasable anthropogenic deposits. Indeed the image of Chernobyl's undigested, unincorporated, unread layers of the fallen leaves of previous years prompts the risky generalization that the Anthropocene might be the name given to the period when technologies of storage and extraction make obsolete the human bearing of tradition.[27]

Poetry is a name for all that houses the remains of a much larger field of symbolic activity—those delicate practices of transmission inseparable from the work of release and abandonment—once and still performed by humans. So by lightly and tenuously holding together the impulse to hold on and the impulse to release and abandon, Bashō's poem, or Hass's translation of it, presents a counterimage to the double logic of compulsive consumption and financial securitization driving the modes of overproduction, which are in no small measure responsible for anthropogenic climate change.

## Notes

1. Tsing's paper is cited in Donna Haraway, "Anthropocene, Capitalocene, Plantationocene, Chthulucene: Making Kin," *Environmental Humanities* 6 (2015): 159–65, 159, http://environmentalhumanities.org/arch/vol6/6.7.pdf.

2. Wes Jackson, *Altars of Unhewn Stone* (New York: North Point, 1987), 140–41 (hereafter cited parenthetically in text).

3. Raymond Williams, *The Country and the City* (Oxford: Oxford University Press, 1975), 102–3.

4. Emily Dickinson, *The Poems of Emily Dickinson*, ed. R. W. Franklin (Cambridge: Belknap, 1998), 1.396 (hereafter cited parenthetically in text). Throughout, I have inserted vertical bars to indicate where the manuscript copy cuts up the implied metrical lines. While, as Annie Finch has argued, the common meter appears to give way and lengthen into two pentameter lines at this poem's end, the drop and isolation of "the Snow" and "the letting go" on the page also seems worth registering; the break at the final "then" indicates a pause that might be indefinite before the gulf to "the letting go" is crossed. See Finch, *The Ghost of Meter: Culture and Prosody in American Free Verse* (Ann Arbor: University of Michigan Press, 1993), 29–30.

5. For example, one version of Dickinson's "It sifts from leaden sieves" ends by figuring this "It" as a decreative artist who erases all traces, even of itself: "Then stills it's Artisans—| like Ghosts—/ Denying they have been—" (1.311).

6. Sharon Cameron, *Lyric Time: Dickinson and the Limits of Genre* (Baltimore: Johns Hopkins University Press, 1979), 168 (hereafter cited parenthetically in text).

7. Earl Miner, *Comparative Poetics: An Intellectual Essay on Theories of Literature* (Princeton: Princeton University Press, 1990), 87–92; Robert Backus, "What Goes into a Haiku," *Literature East and West* 15–16 (1971–72): 735–64, 745; and Dante Gabriel Rossetti, "A Sonnet," http://www.sonnets.org/rossettd.htm.

8. David Archer, *The Long Thaw: How Humans Are Changing the Next 100,000 Years of Earth's Climate* (Princeton: Princeton University Press, 2009), 11 (hereafter cited parenthetically in text).

9. As Ian Welsh writes, "[T]he vigilance required to act as guardians to the longest-lived radioactive isotopes exceeds the duration of all previous human civilisations." *Mobilising Modernity: The Nuclear Moment* (London: Routledge, 2000), 12. See also Barbara Adam's account of the complex temporality and strange materiality of radiation in *Timescapes of Modernity: The Environment and Invisible Hazards* (New York: Routledge, 1998), 10.

10. Relevant here is Srinivas Aravamudan's theorization of what he calls the "catachronism of climate change"—the "inversion of anachronism" by which the present under climate change is shadowed by the advance occurrence of a future whose causes lie in the past. Aravamudan, "The Catachronism of Climate Change," *Diacritics* 41.3 (2013): 6–30, 9. His argument resonates with Barbara Adam's in "The History of the Future," *Rethinking History: The Journal of Theory and Practice* 14.3 (2010): 361–78. Both want to mark the shift or point at which the future that humanist modernity had conceived as "open rather than determined" (Aravamudan, "Catachronism," 9) comes to be recognized as already blocked, defined by industrial technology's "time-space-matter-distantiated outcomes [that] may be latent for hundreds, even thousands of years" (Adam, "History," 373).

11. Aristotle, *Poetics*, 1457b. Derrida discusses the image in "La mythologie blanche" (White mythology), in his *Marges de la philosophie* (Paris: Minuit, 1972), 289–90, but it is Andrzej Warminski who calls it a catachresis in his *Readings in Interpretation: Hölderlin, Hegel, Heidegger* (Minneapolis: University of Minnesota Press, 1987), lvii.

12. Andreas Malm, *Fossil Capital: The Rise of Steam Power and the Roots of Global Warming* (New York: Verso, 2016), 21 (hereafter cited parenthetically in text).

13. Colin Duncan, *The Centrality of Agriculture: Between Humankind and the Rest of Nature* (Montreal: McGill–Queen's University Press, 1996), 46.

14. A reader of Georges Bataille's *The Accursed Share* might say that the problem is with Jackson's metaphor of "coasting," which he applies to both traditional and modern agriculture, saying of the latter that it "coast[s] on the sunlight trapped by floras long extinct." The metaphor inadvertently echoes Malm's

descriptions of the easy take-up and let-down of wind and water. But far from "coasting," concentrated overproduction, Bataille might say, represents a defensive rejection of the Sun's "ceaseless prodigality," an attempt to "stop [its] radiance" by "accumulat[ing] it in growth." Bataille, *The Accursed Share: An Essay on General Economy* (New York: Zone, 1988), 28–29.

15. Jason W. Moore, *Capitalism in the Web of Life: Ecology and the Accumulation of Capital* (New York: Verso, 2015), 43–44 (hereafter cited parenthetically in text).

16. Emily Dickinson, *Poésies complètes*, trans. Françoise Delphy (Paris: Flammarion, 2009).

17. I draw here on my reading of the poem in François, *Open Secrets: The Literature of Uncounted Experience* (Stanford: Stanford University Press, 2008), 202–5.

18. David Porter, *Dickinson: The Modern Idiom* (Cambridge: Harvard University Press, 1981), 26 (hereafter cited parenthetically in text).

19. Ben Panko, "A Radioactive Cold War Military Base Will Soon Emerge from Greenland's Melting Ice," *Smithsonian*, August 5, 2016, http://www.smithsonianmag.com/science-nature/radioactive-cold-war-military-base-will-soon-emerge-greenlands-melting-ice-180960036/?no-ist. Since the article ends with the deflationary admission that "the amount of PCBs and radioactive waste that Camp Century will release is small compared to what already exists in the Arctic," perhaps what this story of melting ice uncovers is nothing material but a monument to the absence of forethought that might also house the strangest of faiths. Was no thought taken of future generations? Or was the waste left as a missive to be unsealed by those who would have used the intervening time to prepare for its receipt?

20. Paul Celan, *Poems*, trans. Michael Hamburger (New York: Persea, 1980), 30–31; the poem was first published in Celan, *Mohn und Gedächtnis* (Stuttgart: Deutsche Verlags-Anstalt, 1952).

21. Among the dead poets' leaves would be Rainer Maria Rilke's "Herbsttag": "Herr: es ist Zeit. Der Sommer war sehr groß. / Leg deinen Schatten auf die Sonnenuhren, / und auf den Fluren laß die Winde los." Rilke, *The Selected Poetry*, trans. Stephen Mitchell (New York: Vintage, 1989), 10.

22. But the male speaker never clearly sides with the past or even rejects the woman's advances. The last line's judgment only puts two nearly identical constatives on the same scale without clarifying their relation: "x is y, you are not [it]" might also be a doubling, "the time is past, you were not on time."

23. Rachel Nuwer, "Forests Around Chernobyl Aren't Decaying Properly," *Smithsonian*, March 14, 2014, http://www.smithsonianmag.com/science-nature/forests-around-chernobyl-arent-decaying-properly-180950075/?no-ist.

24. Robert Hass, *The Essential Haiku: Versions of Bashō, Buson, and Issa* (New York: Ecco, 1994), 15.

25. Roland Barthes, *Le Neutre: Notes de Cours au Collège de France, 1977–1978* (Paris: Seuil, 2002), 148–53.

26. Makoto Ueda, *Bashō and His Interpreters* (Stanford: Stanford University Press, 1992), 112.

27. Needless to say, these deposits will not house the memory of other lost ways of life, whether human or nonhuman. See Shane Gero's opinion piece "The Lost Cultures of Whales," *New York Times*, October 8, 2016, http://www.nytimes.com/2016/10/09/opinion/sunday/the-lost-cultures-of-whales.html?ref=opinion.

# Contributors

*Juliana Chow* is an assistant professor of English at Saint Louis University. Her research in nineteenth-century American literature has appeared in *Arizona Quarterly* and *ESQ*.

*Jeffrey Jerome Cohen* is a professor of English at George Washington University. He is the author of *Of Giants: Sex, Monsters, and the Middle Ages*; *Medieval Identity Machines*; and *Stone: An Ecology of the Inhuman* which won the René Wellek Prize. His numerous edited collections include *Monster Theory: Reading Culture* and *Prismatic Ecology: Ecotheory Beyond Green*.

*Thomas H. Ford* is a lecturer in literary studies at the University of Melbourne. He co-edited (with Tom Bristow) *A Cultural History of Climate Change*.

*Anne-Lise François* is an associate professor in the Departments of English and Comparative Literature at the University of California, Berkeley. She is the author of *Open Secrets: The Literature of Uncounted Experience*, which was awarded the René Wellek Prize.

*Noah Heringman* is the Catherine Paine Middlebush Professor of English at the University of Missouri. He is the author of *Romantic Rocks, Aesthetic Geology* and *Sciences of Antiquity: Romantic Antiquarianism, Natural History, and Knowledge Work*, and the editor of *Romantic Science: The Literary Forms of Natural History*.

*Matt Hooley* is an assistant professor of literature and the environment at Clemson University. He teaches and writes at the intersections of Indigenous, environmental, and American cultural studies and is completing a book tentatively titled "Ordinary Empire: Native Modernism and the Ecologies of Settlement."

*Stephanie LeMenager* is the Moore Endowed Professor of English at the University of Oregon. She is the author of *Living Oil: Petroleum Culture in the American Century* and *Manifest and Other Destinies: Territorial Fictions of the Nineteenth-Century United States*.

*Dana Luciano* is an associate professor of English at Georgetown University. She is the author of *Arranging Grief: Sacred Time and the Body in Nineteenth-Century America* and the co-editor (with Mel Y. Chen) of "Queer Inhumanisms" (a special issue of *GLQ*) and (with Ivy G. Wilson) of *Unsettled States: Nineteenth-Century American Literary Studies*.

*Tobias Menely* is an associate professor of English at the University of California, Davis. He is the author of *The Animal Claim: Sensibility and the Creaturely Voice* and is currently writing "The Climatological Unconscious."

*Steve Mentz* is a professor of English at St. John's University in New York City. He is the author of *Shipwreck Modernity: Ecologies of Globalization, 1550–1719*; *At the Bottom of Shakespeare's Ocean*; and *Romance for Sale in Early Modern England*.

*Benjamin Morgan* is an associate professor of English at the University of Chicago. He is the author of *The Outward Mind: Materialist Aesthetics in Victorian Science and Literature*.

*Justin Neuman*, a scholar of twentieth- and twenty-first-century Anglophone literature and culture, teaches at the New School. He is the author of *Fiction Beyond Secularism*.

*Jesse Oak Taylor* is an associate professor of English at the University of Washington. He is the author of *The Sky of Our Manufacture: The London Fog in British Fiction from Dickens to Woolf* which won the ASLE Book Award, and co-author (with Daniel C. and Carl E. Taylor, of Empowerment on an Unstable Planet: From Seeds of Human Energy to a Scale of Global Change).

*Jennifer Wenzel* is an associate professor in the Department of English and Comparative Literature and the Department of Middle Eastern, South Asian, and African Studies at Columbia University. She is the author of *Bulletproof: Afterlives of Anticolonial Prophecy in South Africa and Beyond* and a co-editor (with Imre Szeman and Patricia Yaeger) of *Fueling Culture: 101 Words for Energy and Environment*.

*Derek Woods* is a postdoctoral fellow in the Dartmouth Society of Fellows. His work appears in *American Literary History*, *Victorian Literature and Culture*, and the *minnesota review*. He is at work on two book projects, "What is Ecotechnology?" and "The Poetics of Scale."

# Index

## A

Adams, Henry, 17, 151–64
aesthetics, 69, 78, 80–83, 85–86, 97–98, 126, 128, 131 n. 16, 133, 135–37, 140, 144–47, 155, 157, 164, 216
agriculture, 14, 26–27, 31, 45, 47, 128, 245–46
Alaimo, Stacy, 41 n. 20, 213–15, 221
Anthropocene (defined), 2–10
  critiques of, 3, 8–9
  periodization and, 8, 14–15, 18–20, 23 n. 39, 26–28, 32–38, 43–48, 54, 56, 83, 91–93, 122, 137, 202–4, 207
  1610 (Orbis Hypothesis), 4, 14, 16, 19, 26, 34, 37, 39, 43–56, 113, 191
  1784/1800 (Industrial Revolution), 3–6, 14, 50, 62, 78–95, 158, 162, 191, 205, 245
  1945/1950 (nuclear test/ Great Acceleration), 3, 6, 8, 14, 18, 20, 26, 32, 45, 47–48, 50, 56, 137, 151, 191, 202–10, 213
Anthropocene Working Group, 2–3, 6, 8–10, 41 n. 20, 64–65, 109, 113, 135, 204
apocalypse, 68, 103, 125, 225–28, 233
Archer, David, 244, 248, 250, 254
Aravamudan, Srinivas, 162, 257 n. 10
Aristotle, 44, 62, 137, 244–45
atmosphere, 4, 15–16, 26, 37–38, 44, 48, 79, 83, 85–89, 93–94

## B

Babbage, Charles, 105–12, 115 n. 32
Bacigalupi, Paolo, 71, 234
Bashō, Matsuo, 254–56
Baucom, Ian, 29, 147, 179–80
Benjamin, Walter, 1, 12, 21, 24 n. 41, 180–81, 182 n. 26
Bennett, Jane, 28, 112–13
Best, Stephen, 11, 63, 175–76
biodiversity, 122, 202–4, 208, 213–16, 235
biopolitics, 88, 194, 213
biosphere, 4, 9, 15–16, 161, 203–4, 208, 210, 212, 214–16, 218 n. 27
Brontë, Charlotte, 14, 79, 85–94
Buckland, William, 96, 107
Buffon, Georges-Louis Leclerc, Comte de, 5, 17–18, 59–74
Burnet, Thomas, 2, 66, 103
Butler, Octavia, 209, 223–37

## C

Cameron, Sharon, 117, 242–43
Celan, Paul, 252, 254–55
Chakrabarty, Dipesh, 4, 68, 120, 132, 141, 146, 167–73, 187–93, 230, 243
coal. *See* energy, fossil fuels
Cohen, Jeffrey Jerome, 2, 15, 18–20, 46, 65, 207
Colebrook, Claire, 61–62, 67–68
cities, 155, 157, 161, 177–78, 213, 220, 232–33, 250
climate
  climate change, 4, 6–7, 15–19, 29–30, 36–39, 48, 70–71, 86, 93, 109, 117, 130 n. 1, 135, 140–41, 146–47, 150–54, 160–64, 171–73, 185, 187–89, 215–16, 220–36, 239, 241, 244, 249–51, 256
  Intergovernmental Panel on (IPCC), 162, 206

climate (*continued*)
    climate refugees, 19, 21, 28, 39, 191, 224
    and/as history, 29, 84–86
Coates, Ta-Nehisi, 225–29
Coetzee, J. M., 55, 167–81
colonialism, 49, 53–54, 70, 88, 100, 116 n. 40, 171–74, 179–80, 184–99, 200 nn. 7, 16, 214, 224–30, 234–36
Crutzen, Paul, 3, 5, 119, 161, 191
cryosphere, 15. *See also* ice cores
Cuvier, Georges, 14, 23 n. 25, 60–63, 66, 68–72, 97, 103

## D

Darwin, Charles, 118–21, 128, 134–35, 142–43, 146, 148 n. 17, 208–9, 213
Derrida, Jacques, 7, 100, 213, 257 n. 11
Dickinson, Emily, 241–52
dystopia, 62, 76 n. 39, 220, 224. *See also* utopia

## E

empire, 34, 167–81, 187, 190, 201 n. 22. *See also* colonialism and imperialism
energy, 4–5, 12, 17–18, 20, 71–72, 106, 108, 132, 147, 150–64, 204–7, 240–41, 244–49
    fossil fuels, 3, 5, 16–18, 21, 60–63, 71–74, 76 n. 31, 152–53, 156, 160–62, 173, 202, 204, 207, 215, 225, 241, 244–45, 249
        coal, 5, 45, 47, 50, 60, 62–65, 71–73, 151–56, 159–63, 207, 245–48
        petroleum, 31, 47, 50, 150–53, 159–60, 185, 191, 203, 207, 211, 217 n. 11
    nuclear, 47, 159, 244
    renewable, 21
    solar, 244–45
Earth system, 2, 4–14, 18, 37–38, 48, 56, 66, 73, 79, 84, 109, 119–22, 153, 186, 202–6, 215, 244. *See also* atmosphere and biosphere and cryosphere and hydrosphere and lithosphere
ethics, 27, 132. *See also* justice
extinction, 6, 20, 23 n. 25, 26–27, 31, 59–64, 66–74, 77 n. 44, 96–97, 101, 103, 106, 111, 122, 137, 145, 172–73, 178–79, 202–9, 212, 214–15

## F

forest(s), 37, 49, 52, 65, 72, 118, 120–22, 129, 225, 243, 245, 248–49, 254
    deforestation, 26–27, 31, 153
form (literary), 2, 11–15, 18–20, 26–27, 44–46, 55–56, 60, 62–63, 73–74, 79, 83, 92, 117–18, 125–27, 132–47, 194, 199, 204, 209–10, 220, 222–23
    formalism, 11–12, 132–41
    life form, 203, 206, 208, 210, 212–16
    *See also* unconformity
fossils, 6, 13, 15–6, 23 n. 25, 25–26, 40 n. 3, 44, 59–63, 67, 70–73, 77 n. 44, 96–114, 121–22, 126, 142–43, 152, 170, 194, 213
fossil fuel. *See* energy, fossil fuels
Freud, Sigmund, 175–78

## G

geo-engineering, 17, 21, 72, 163
gender, 88, 105, 114 n. 6, 118, 226, 231–34, 252, 258 n. 22
genre, 1, 13–14, 18, 20, 27, 33, 44–46, 51, 54–55, 112, 133, 137, 143, 177–78, 220–23, 234
    cli-fi, 71, 221–23, 228, 234
    evolution and, 203, 209
    extinction and, 68
    lyric, 18, 33, 80, 85, 239–56
    medieval, 28, 33, 35, 65, 75 n. 6
    novel, 1, 13, 18–20, 64, 71, 79–94, 140–41, 144–47, 152, 220, 223–25, 227–28, 235–36
    poetics, 12–15, 33–35, 49, 59, 66, 79, 80–87, 99, 101–5, 112, 136–39, 144, 157, 186–99, 216, 242–43, 248–56
    romance, 2, 13, 17–18, 22 n. 5, 27, 59, 64, 66, 133, 141, 177–78, 233
    scientific, 28, 59–63, 69, 71–74
geology, 2, 5, 8–9, 14, 18, 30, 33, 59–74, 97–105, 110–11, 132–40, 171–81, 194. *See also* stratigraphy
Gidal, Eric, 15, 24 n. 40, 33–35, 180

## H

Hamilton, Clive, 7, 61, 64–65, 73, 75 n. 5, 138
Haraway, Donna, 8, 118, 205, 214, 226–28, 234

Hardy, Thomas, 18, 55, 132–47, 209
Hass, Robert, 254–56
Hegel, G. W. F., 80, 83
Heise, Ursula, 146, 202
Heringman, Noah, 17–18, 20, 25–26, 97–98, 105
Hitchcock, Edward, 96–114
Holocene, 6, 9–10, 31–32, 48, 64, 172, 216 n. 1, 217 n. 9, 234–35, 239
Hutton, James, 5, 30, 97, 103
hydrosphere, 4, 180

I

ice cores, 4, 12, 15, 24, 30, 43–44, 91, 113
imperialism, 19, 162, 174. *See also* colonialism and empire
Indigenous, 12, 16, 34, 37, 111–13, 180, 184–99

J

Jackson, Wes, 240, 246
Jameson, Fredric, 11, 63, 224, 237 n. 14
justice, 12, 15, 56, 112, 120, 168, 171, 173, 179, 181, 230, 249

K

Kant, Immanuel, 79, 81–82, 85–86, 145–46, 202
Kingsolver, Barbara, 235
Kolbert, Elizabeth, 17, 59–62, 66–74, 109

L

Latour, Bruno, 2, 4, 51, 162
LeMenager, Stephanie, 12, 16, 19–21, 170, 217 n. 7
Levine, Caroline, 11, 45, 136–37, 139–40
Lewis, Simon L., 3–4, 8, 16, 22 n. 21, 37, 43–44, 48–49, 53, 63–65, 67–68, 74, 75 n. 5, 113, 116 n. 40, 177, 182 n. 14, 187
lithosphere, 4, 15, 31, 180
Luciano, Dana, 16, 20, 49, 189
Lyell, Charles, 14, 21–22 n. 5, 96, 100, 121, 134, 142, 167, 171–73, 177–80

M

Malm, Andreas, 8, 245–47
Mandel, Emily St. John, 231–33
Marcus, Sharon, 11, 63, 175–76
Marx, Karl, 247
Marsh, George Perkins, 5, 19, 119–20, 130 n. 9
Maslin, Mark A., 3–4, 8, 16, 22 n. 21, 37, 43–44, 48–49, 53, 63–65, 67–68, 74, 75 n. 5, 113, 116 n. 40, 177, 182 n. 14, 187
Mbembe, Achille, 177–78, 187, 190–91
McCarthy, Cormac, 230–31
McKibben, Bill, 6, 162
McLuhan, Marshall, 206–7
media, 15, 20, 27, 44, 140, 156, 170, 202, 204–7, 222, 231
modernity, 1, 80, 102, 163, 176, 221–23, 232–33, 241, 257 n. 10
  early modern, 33–38, 45, 47–53, 62
  modernism, 17, 23 n. 39, 151–54, 157, 210
  modern constitution, 4, 21
  postmodern, 50–51, 55
Moore, Jason, 4, 8, 50, 247
Moretti, Franco, 11, 203, 209–10, 214, 223
Morton, Timothy, 28, 140, 146
Mumford, Lewis, 155–56, 162

N

narrative
  Anthropocene as, 3, 5, 12, 26, 46, 55–56, 66–67, 138
  geology and, 21 n. 5, 22 n. 7, 64–65, 135
  limitations of, 29
  periodization and. *See* periodization
  scale and, 1–2, 67, 132–33, 140, 143, 145
  visual, 232, 236
  *See also* form, novel, and genre
necropolitics, 185–98
New Materialism, 11–12, 97–99, 109–13, 114 n. 6, 116 n. 39, 135–36
Ngai, Sianne, 81–82
Nixon, Rob, 12, 202, 225
nuclear
  power, 159, 244
  trace, *see* Anthropocene, nuclear test / Great Acceleration
  winter/apocalypse, 150, 164 n. 2, 231
Noah (Biblical), 28, 30, 34–39

## O

okpik, d. g. nanouk, 184–99

## P

Parikka, Jussi, 116 n. 39, 207
population, 31, 49, 51, 65, 132, 155, 162, 204, 216 n. 4, 234

## R

racism, 188–90, 227
Raup, David, 208–10, 210–15
reading
  close, 13, 26, 30–38, 88, 112, 120, 224
  distant, 26, 37–38, 209–10, 214, 222
  surface, 11–12, 15, 26–27, 38, 175–76, 181
  symptomatic, 11–13, 63, 141, 145, 175, 206, 221
  stratigraphy and/as, 2–12, 16, 35–36, 44, 60, 67, 85, 125, 137, 171, 174–75, 180, 191
Rich, Nathaniel, 233
Robinson, Kim Stanley, 146, 230
Romanticism, 17, 20, 78–94, 98, 105, 251
romance. *See* genre, romance
Rudwick, Martin, 5, 23 n. 25, 63, 72, 75 n. 6, 133

## S

scale, 12, 18–19, 27, 48–49, 59, 62, 67, 73, 83, 110, 117, 119–20, 122, 132–47, 150, 153, 161–62, 173, 184–99, 202, 214, 235
  geologic time scale (GTS), 3, 6, 9, 60, 63, 135, 137–40, 225, 230
  individual, 19, 177
  scalability, 19, 187–99
  scalar shifts, 6, 133
  scale variance, 14, 18, 73, 120, 132, 140–47, 179, 188, 196, 209–10, 243
  spatial, 2, 78, 119–22, 189, 203
  temporal, 1–2, 11, 25, 30, 61–62, 68, 72, 103, 145, 154, 169, 171, 175, 177, 192, 222, 243–44
Scranton, Roy, 189–91, 227–31, 235–36
sexuality, 36, 114 n. 6, 233–34, 253

slavery, 49–50, 54, 110–11, 115–116 n. 32, 190, 199, 200 n. 16, 224
species, 20, 27, 31, 36, 59–62, 67–73, 96–97, 103, 111, 118, 120–24, 134, 146, 172–73, 178–79, 184–88, 194, 199, 202–16, 227–30, 239–40, 246, 251
  human as, 2–3, 6, 8–9, 12, 21, 25, 51, 59–61, 65, 67–70, 73, 120, 132, 160, 163, 173, 192, 230, 233, 244
  *See also* extinction and biodiversity
Shakespeare, William, 36–37, 47–56, 104, 209, 232
Shelley, Percy Bysshe, 79, 85–86
Shelley, Mary, 66
Snow, C. P., 114
Stoermer, Eugene, 3, 119, 191
stratigraphy, 9–10, 30–35, 101, 109, 139, 170–75, 177
  International Commission on, 31, 109, 138
  *See also* geology and Earth systems and reading, stratigraphy as
Szerszynski, Bronislaw, 9, 40 n. 3

## T

technology, 5–7, 16–17, 21, 32, 38, 53, 64, 78, 81–83, 152, 155–63, 195, 207, 236, 256
technics, 155, 162
technofossils, 6, 67
technosphere, 204
techno-utopianism, 17
Theroux, Marcel, 231–33
Thoreau, Henry David, 117–29
Tsing, Anna, 8, 187, 196–97, 239

## U

unconformity, 13, 15–18, 24, 30–31, 33–38, 41, 180
utopia, 17, 157–60, 224, 237 n. 14, 238 n. 23, 243. *See also* dystopia

## V

VanderMeer, Jeff, 235
Vernadsky, Vladimir, 203, 215, 218 n. 27
Vogt, E. van, 206–8

## W

Wallerstein, Immanuel, 4
Weisman, Alan, 167, 178
Wells, H. G., 60–61, 67
Williams, Raymond, 85, 212, 240

## Y

Yaeger, Patricia, 147, 152

## Z

Zalasiewicz, Jan, 2, 17–18, 37, 59–67, 74, 109, 135, 137, 167–68, 171, 173, 178–80
Zola, Emile, 155